ADAPTATION IN NATURAL AND ARTIFICIAL SYSTEMS

Complex Adaptive Systems

John H. Holland, Christopher Langton, and Stewart W. Wilson, advisors

Adaptation in Natural and Artificial Systems: An Introductory Analysis with Applications to Biology, Control, and Artificial Intelligence, MIT Press/ Bradford Books edition
John H. Holland

ADAPTATION IN NATURAL AND ARTIFICIAL SYSTEMS

An Introductory Analysis with Applications to Biology, Control, and Artificial Intelligence

John H. Holland

A Bradford Book
The MIT Press
Cambridge, Massachusetts
London, England

Fifth printing, 1998
First MIT Press edition, 1992

This book was printed and bound in the United States of America.

Library of Congress Cataloging-in-Publication Data

Holland, John H. (John Holland), 1929–
Adaptation in natural and artificial systems : an introductory analysis with applications to biology, control, and artificial intelligence / John H. Holland. — 1st MIT Press ed.
p. cm. — (Complex adaptive systems)
Includes bibliographical references and index.
ISBN 0-262-08213-6 (hc). — ISBN 0-262-58111-6 (pbk.)
1. Adaptation (Biology)—Mathematical models. 2. Adaptive control systems—Mathematical models. I. Title. II. Series.
QH546.H64 1992
574.5′01′5118—dc20 91-41044
CIP

Contents

List of Figures

Figure

Preface to the 1992 Edition

When this book was originally published I was *very* optimistic, envisioning extensive reviews and a kind of "best seller" in the realm of monographs. Alas! That did not happen. After five years I did regain some optimism because the book did not "die," as is usual with monographs, but kept on selling at 100–200 copies a year. Still, research in the area was confined almost entirely to my students and their colleagues, and it did not fit into anyone's categories. "It is certainly *not* a part of artificial intelligence" and "Why would somebody study learning by imitating a process that takes billions of years?" are typical of comments made by those less inclined to look at the work.

Five more years saw the beginnings of a rapid increase in interest. Partly, this interest resulted from a change of focus in artificial intelligence. Learning, after several decades at the periphery of artificial intelligence, was again regarded as pivotal in the study of intelligence. A more important factor, I think, was an increasing recognition that genetic algorithms provide a tool in areas that do not yield readily to standard approaches. Comparative studies began to appear, pointing up the usefulness of genetic algorithms in areas ranging from the design of integrated circuits and communication networks to the design of stock market portfolios and aircraft turbines. Finally, and quite important for future studies, genetic algorithms began to be seen as a theoretical tool for investigating the phenomena generated by *complex adaptive systems*—a collective designation for nonlinear systems defined by the interaction of large numbers of adaptive agents (economies, political systems, ecologies, immune systems, developing embryos, brains, and the like).

The last five years have seen the number of researchers studying genetic algorithms increase from dozens to hundreds. There are two recent innovations that will strongly affect these studies. The first is the increasing availability of massively parallel machines. Genetic algorithms work with populations, so they are intrinsically suited to execution on computers with large numbers of processors, using a processor for each individual in the population. The second innovation is a unique interdisciplinary consortium, the Santa Fe Institute, dedicated to the study of complex adaptive systems. The Santa Fe Institute, by providing a focus for intensive interactions among its collection of Nobel Laureates, MacArthur Fellows, Old and Young Turks, and bright young postdocs, has already made a substantial impact in the field of economics. Current work emanating from the Institute promises similar effects in fields ranging from studies of the immune system to studies of new approaches to cognitive science. The future for studies of adaptive systems looks bright indeed.

Fifteen years should provide perspective and a certain detachment. Despite that, or because of it, I still find the 1975 preface surprisingly relevant. About the only change I would make would be to put more emphasis on improvement and less on optimization. Work on the more complex adaptive systems—ecologies, for example—has convinced me that their behavior is not well described by the trajectories around global optima. Even when a relevant global optimum can be defined, the system is typically so "far away" from that optimum that basins of attraction, fixed points, and the other apparatus used in studying optima tell little about the system's behavior. Instead, competition between components of the system, aimed at "getting an edge" over neighboring competitors, determines the aggregate behavior. In all other respects, I would hold to the points made in the earlier preface.

There are changes in emphasis reflected by two changes in terminology since 1975. Soon after the book was published, doctoral students in Ann Arbor began using the term *genetic algorithm* in place of *genetic plan*, emphasizing the centrality of computation in defining and implementing the plans. More recently, I've advocated *implicit parallelism* over *intrinsic parallelism* to distinguish the "implicit" workings of the algorithm, via schemata, from the parallel processing of the populations used by the algorithm.

As a way of detailing some more recent ideas and research, I've added a new chapter, chapter 10, to this edition. In part, this chapter concerns itself with further work on the advanced questions posed in section 9.3 of the previous edition. Questions concerning the design of systems that build experience-based, hierarchical models of their environments are addressed in section 10.1 of the new chapter. Questions concerning speciation and the evolution of ecologies are addressed in terms of the *Echo* models in section 10.3. The Echo models, besides being concerned with computer-based gedanken experiments on these questions, have a broader purpose. They are designed to facilitate investigation of mechanisms, such as competition and trading, found in a wide range of complex adaptive systems. In addition to these discussions, the new chapter also includes, in section 10.2, some corrections to the original edition. Section 10.4 concludes the chapter with a new set of advanced questions and some new speculations.

There is more recent work, still in its earliest stages, that is not discussed in chapter 10. Freddy Christiannsen, Marc Feldman, and I, working through the Santa Fe Institute, have begun to introduce the effects of schemata into much-generalized versions of Fisher's equations. This work is, in part, a follow-up of work of a decade ago, by Bob Axelrod and Bill Hamilton, that began to study the relation of recombination to the prevalence of sex in spite of the two-fold genetic load it incurs. In another direction, some preliminary theoretical investigations, stimulated by the Echo models, suggest that there is a schema theorem that is relevant to any adaptive system that can be described in terms of resource flow—such a system may involve neither reproduction nor defined fitness functions. Also in the wings is a characterization of a broad class of problems or "landscapes" that are relatively easy for genetic algorithms but difficult for both traditional optimization techniques and new weight-changing techniques such as artificial nerve nets and simulated annealing.

At a metalevel, the problem landscapes we've been studying may describe an essential aspect of all problems encountered by complex adaptive systems: Easy (linear, hill-climbing) problems have a short-lived influence on complex adaptive systems because they are quickly exploited and absorbed into the system's structure. Extremely difficult problems ("spikes") almost never influence the behavior because they are almost never solved. This leaves as a major continuing influence the presence of certain kinds of "bottlenecks." These bottlenecks are regions in the problem space that offer improvement but are surrounded by "valleys" of lowered performance. The time it takes to traverse these valleys determines the trajectory, and rate of improvement, of the adaptive system. It seems likely that this rate will be determined, to a great degree, by recombination applied to "building blocks" (schemata) supplied by solutions attached to other regions of high performance.

It is an exciting time to study adaptation in natural and artificial systems; perhaps these studies will yield another edition sometime in the next millenium.

John H. Holland
October 1991

Preface

The first technical descriptions and definitions of adaptation come from biology. In that context adaptation designates any process whereby a structure is progressively modified to give better performance in its environment. The structures may range from a protein molecule to a horse's foot or a human brain or, even, to an interacting group of organisms such as the wildlife of the African veldt. Defined more generally, adaptive processes have a critical role in fields as diverse as psychology ("learning"), economics ("optimal planning"), control, artificial intelligence, computational mathematics and sampling ("statistical inference"). Basically, adaptive processes are optimization processes, but it is difficult to subject them to unified study because the structures being modified are complex and their performance is uncertain. Frequently nonadditive interaction (i.e., "epistasis" or "nonlinearity") makes it impossible to determine the performance of a structure from a study of its isolated parts. Moreover possibilities for improved performance must usually be exploited at the same time that the search for further improvements is pressed. While these difficulties pose a real problem for the analyst, we know that they are routinely handled by biological adaptive processes, *qua* processes. The approach of this book is to set up a mathematical framework which makes it possible to extract and generalize critical factors of the biological processes. Two of the most important generalizations are: (1) the concept of a *schema* as a generalization of an interacting, coadapted set of genes, and (2) the generalization of genetic operators such as crossing-over, inversion, and mutation. The schema concept makes it possible to dissect and analyze complex "nonlinear" or "epistatic" interactions, while the generalized genetic operators extend the analysis to studies of learning, optimal planning, etc. The possibility of "intrinsic parallelism"—the testing of many schemata by testing a single structure—is a direct offshoot of this approach. The book develops an extensive study of intrinsically parallel processes and illustrates their uses over the full range of adaptive processes, both as hypotheses and as algorithms.

The book is written on the assumption that the reader has a familiarity with probability and combinatorics at the level of a first course in finite mathematical structures, plus enough familiarity with the concept of a system to make the notion of "state" a comfortable working tool. Readers so prepared should probably read the book in the given order, moving rapidly (on the first reading) over any example or proof offering more than minor difficulties. A good deal of meaning can still be extracted by those with less mathematics if they are willing to abide the notation, treating the symbols (with the help of the Glossary) as abbreviations for familiar intuitive concepts. For such a reading I would recommend chapter 1 (skipping over section 1.3), the first part of chapter 4 and the summary at the end, the discussions throughout chapter 6 (particularly section 6.6), section 7.5, and most of chapter 9, combined with use of the Index to locate familiar examples and topics. The reader whose first interest is the mathematical development (exclusive of applications) will find section 2.2, chapters 4 and 5, sections 6.2, 6.3, 6.4, 7.1, 7.2, 7.3, 7.5, and 9.1 the core of the book. By a judicious use of the Glossary and Index it should be possible for a well-trained system scientist to tackle this part of the book directly. (This is *not* a procedure I would recommend except as a way of getting a mathematical overview before further reading; in a book of this sort the examples have a particularly important role in establishing the *meaning* of the formalism.)

The pattern of this book, as the reader sees it now, only distantly resembles the one projected at its inception. The first serious writing began almost seven years ago at Pohoiki on the Big Island under the kamaaina hospitality of Carolyn and Gilbert Hay. No book could start in a finer setting. Since that time whole chapters, including chapters on hierarchies, the Kuhn-Tucker fixed point theorem, and cellular automata, have come and gone, a vital $\mu_c(f)$ emerged, blossomed and disappeared, 2-armed bandits arrived, and so on. At this remove it would be about as difficult to chronicle those changes as to acknowledge properly the people who have influenced the book along the way. Arthur Burks stands first among those who provided the research setting and encouragement which made the book feasible; Michael Arbib's comments on a near-final draft stand as the culmination of readings, written comments, commentaries, and remarks by more than a hundred students and colleagues; and Monna Whipp's perseverance through the typing of the final draft and revised revisions of changes brings to fruition the tedious work of her predecessors. For the rest, I cannot conceive that appearance in a long list of names is a suitable reward, but I also cannot conceive a good alternative (beyond personal expression), so they remain anonymous and bereft of formal gratitude beyond some appearances in the references. They deserve better.

JOHN H. HOLLAND

ADAPTATION IN NATURAL AND ARTIFICIAL SYSTEMS

1. The General Setting

1. INTRODUCTION

How does evolution produce increasingly fit organisms in environments which are highly uncertain for individual organisms?

What kinds of economic plan can upgrade an economy's performance in spite of the fact that relevant economic data and utility measures must be obtained as the economy develops?

How does an organism use its experience to modify its behavior in beneficial ways (i.e., how does it learn or "adapt under sensory guidance")?

How can computers be programmed so that problem-solving capabilities are built up by specifying "*what* is to be done" rather than "*how* to do it" ?

What control procedures can improve the efficiency of an ongoing process, when details of changing component interactions must be compiled and used concurrently?

Though these questions come from very different areas, it is striking how much they have in common. Each involves a problem of optimization made difficult by substantial complexity and uncertainty. The complexity makes discovery of the optimum a long, perhaps never-to-be-completed task, so the best among *tested* options must be exploited at every step. At the same time, the uncertainties must be reduced rapidly, so that knowledge of *available* options increases rapidly. More succinctly, information must be exploited as acquired so that performance improves apace. Problems with these characteristics are even more pervasive than the questions above would indicate. They occur at critical points in fields as diverse as evolution, ecology, psychology, economic planning, control, artificial intelligence, computational mathematics, sampling, and inference.

There is no collective name for such problems, but whenever the term *adaptation* (*ad* + *aptare*, to fit to) appears it consistently singles out the problems of interest. In this book the meaning of "adaptation" will be extended to encompass the entire collection of problems. (Among rigorous studies of adaptation, Tsypkin's [1971] usage comes closest to this in breadth, but he deliberately focuses on the man-made systems.) This extension, if taken seriously, entails a commitment to view adaptation as a fundamental process, appearing in a variety of guises but subject to unified study. Even at the outset there is a powerful warrant for this view. It comes from the observation that all variations of the problem give rise to the same fundamental questions.

> To what parts of its environment is the organism (system, organization) adapting?
> How does the environment act upon the adapting organism (system, organization)?
> What structures are undergoing adaptation?
> What are the mechanisms of adaptation?
> What part of the history of its interaction with the environment does the organism (system, organization) retain?
> What limits are there to the adaptive process?
> How are different (hypotheses about) adaptive processes to be compared?

Moreover, as we attempt to answer these questions in different contexts, essentially the same obstacles to adaptation appear again and again. They appear with different guises and names, but they have the same basic structure. For example, "nonlinearity," "false peak," and "epistatic effect" all designate versions of the same difficulty. In the next section we will look more closely at these obstacles; for now let it be noted that the study of adaptation is deeply concerned with the means of overcoming these obstacles.

Despite a wealth of data from many different fields and despite many insights, we are still a long way from a general understanding of adaptive mechanisms. The situation is much like that in the old tale of blind men examining an elephant—different aspects of adaptation acquire different emphases because of the points of contact. A specific feature will be prominent in one study, obscure in another. Useful and suggestive results remain in comparative isolation. Under such circumstances theory can be a powerful aid. Successful analysis separates incidental or "local" exaggerations from fundamental features. A broadly conceived analytic theory brings data and explanation into a coherent whole, providing opportunities for prediction and control. Indeed there is an important sense in which a good theory defines the objects with which it deals. It reveals their interactions, the methods of transforming and controlling them, and predictions of what will happen to them.

Theory will have a central role in all that follows, but only insofar as it illuminates practice. For natural systems, this means that theory must provide techniques for prediction and control; for artificial systems, it must provide practical algorithms and strategies. Theory should help us to know more of the mechanisms of adaptation and of the conditions under which new adaptations arise. It should enable us to better understand the processes whereby an initially unorganized system acquires increasing self-control in complex environments. It should suggest procedures whereby actions acquired in one set of circumstances can be transferred to new circumstances. In short, theory should provide us with means of prediction and control not directly suggested by compilations of data or simple tinkering. The development here will be guided accordingly.

The fundamental questions listed above can serve as a starting point for a unified theory of adaptation, but the informal phrasing is a source of difficulty. With the given phrasing it is difficult to conceive of answers which would apply *unambiguously* to the full range of problems. Our first task, then, is to rephrase questions in a way which avoids ambiguity and encourages generality. We can avoid ambiguity by giving precise definitions to the terms appearing in the questions, and we can assure the desired generality if the terms are defined by embedding them in a common formal framework. Working within such a framework we can proceed with theoretical constructions which are of real help in answering the questions. This, in broad outline, is the approach we will take.

2. PRELIMINARY SURVEY

(Since we are operating outside of a formal framework in this chapter, some of the statements which follow will be susceptible of different, possibly conflicting interpretations. Precise versions will be formulated later.)

Just what are adaptation's salient features? We can see at once that adaptation, whatever its context, involves a progressive modification of some structure or structures. These structures constitute the grist of the adaptive process, being largely determined by the field of study. Careful observation of successive structural modifications generally reveals a basic set of structural modifiers or operators; repeated action of these operators yields the observed modification sequences. Table 1 presents a list of some typical structures along with the associated operators for several fields of interest.

A system undergoing adaptation is largely characterized by the mixture of operators acting on the structures at each stage. The set of factors controlling this changing mixture—the adaptive plan—constitutes the works of the system as far as its adaptive character is concerned. The adaptive plan determines just what structures arise in response to the environment, and the set of structures attainable

Table 1: Typical Structures and Operators

Field	*Structures*	*Operators*
Genetics	chromosomes	mutation, recombination, etc.
Economic planning	mixes of goods	production activities
Control	policies	Bayes's rule, successive approximation, etc.
Physiological psychology	cell assemblies	synapse modification
Game theory	strategies	rules for iterative approximation of optimal strategy
Artificial intelligence	programs	"learning" rules

by applying all possible operator sequences marks out the limits of the adaptive plan's domain of action. Since a given structure performs differently in different environments—the structure is more or less fit—it is the adaptive plan's task to produce structures which perform "well" (are fit) in the environment confronting it. "Adaptations" to the environment are persistent properties of the sequence of structures generated by the adaptive plan.

A precise statement of the adaptive plan's task serves as a key to uniform treatment. Three major components are associated in the task statement: (1) the environment, E, of the system undergoing adaptation, (2) the adaptive plan, τ, whereby the system's structure is modified to effect improvements, (3) a measure, μ, of performance, i.e., the fitness of the structures for the environment. (The formal framework developed in chapter 2 is built around these three components.) The crux of the problem for the plan τ is that initially it has incomplete information about which structures are most fit. To reduce this uncertainty the plan must test the performance of different structures in the environment. The "adaptiveness" of the plan enters when different environments cause different sequences of structures to be generated and tested.

In more detail and somewhat more formally: A characteristic of the environment can be unknown (from the adaptive plan's point of view) only if alternative outcomes of the plan's tests are allowed for. Each distinct combination of alternatives is a distinct environment E in which the plan may have to act. The set of all possible combinations of alternatives indicates the plan's initial uncertainty about the environment confronting it—the range of environments in which the plan should be able to act. This initial uncertainty about the environment will be formalized by designating a class $\mathcal{E}$ of possible environments. The domain of action of the adaptive plan will be formalized by designating a set $\mathcal{A}$ of attainable structures. The fact that different $E \in \mathcal{E}$ in general elicit different performances from a given structure $A \in \mathcal{A}$ means formally that there will be a different performance measure μ_E associated with each E. Each field of study is typified as much by its performance measures as by its structures and operators. For the fields mentioned in connection with examples of structures and operators, we have a corresponding list of performance measures:

Table 2: Typical Performance Measures

Field	*Performance Measure*
Genetics	Fitness
Economic planning	Utility
Control	Error functions
Physiological psychology	Performance rate (in some contexts, but often unspecified)
Game theory	Payoff
Artificial intelligence	Comparative efficiency (if specified at all)

The successive structural modifications dictated by a plan τ amount to a sequence or trajectory through the set $\mathcal{A}$. For the plan to be adaptive the trajectory through $\mathcal{A}$ must depend upon which environment $E \in \mathcal{E}$ is present. Symbolizing the set of operators by Ω, this last can be stated another way by saying that the order of application of operators from Ω must depend upon E.

It is clear that the organization of $\mathcal{A}$, the effects of the operators Ω upon structures in $\mathcal{A}$, and the form of the performance measure μ_E all affect the difficulty of adaptation. Among the specific obstacles confronting an adaptive plan are the following:

1. $\mathcal{A}$ is large so that there are many alternatives to be tested.
2. The structures in $\mathcal{A}$ are complicated so that it is difficult to determine which substructures or components (if any) are responsible for good performance.
3. The performance measure μ_E is a complicated function with many interdependent parameters (e.g., it has many dimensions and is nonlinear, exhibiting local optima, discontinuities, etc.).

4. The performance measure varies over time and space so that given adaptations are only advantageous at certain places and times.
5. The environment E presents to τ a great flux of information (including performances) which must be filtered and sorted for relevance.

By describing these obstacles, and the adaptive plans meant to overcome them, within a general framework, we open the possibility of discovering plans useful in any situation requiring adaptation.

Before going further let us flesh out these abstractions by using them in the description of two distinct adaptive systems, one simple and artificial, the other complex and natural.

3. A SIMPLE ARTIFICIAL ADAPTIVE SYSTEM

The artificial adaptive system of this example is a pattern recognition device. (The device to be described has very limited capabilities; while this is important in applications, it does not detract from the device's usefulness as an illustration.) The information to be fed to the adaptive device is preprocessed by a rectangular array of sensors, a units high by b units wide. Each sensor is a threshold device which is activated when the light falling upon it exceeds a fixed threshold. Thus, when a "scene" is presented to the sensor array at some time t, each individual sensor is either "on" or "off" depending upon the amount of light reaching it. Let the activity of the ith sensor, $i = 1, 2, \ldots, ab$, at time t be represented formally by the function $\delta_i(t)$, where $\delta_i(t) = 1$ if the sensor is "on" and $\delta_i(t) = 0$ if it is "off." A given scene thus gives rise to a configuration of ab "ones" and "zeros." All told there are 2^{ab} possible configurations of sensor activation; let C designate this set of possible configurations. It will be assumed that a particular subset of C_1 of C corresponds to (instances of) the pattern to be recognized. The particular subset involved, among the $2^{2^{ab}}$ possible, will be unknown to the adaptive device. (E.g., C_1 might consist of all configurations containing a connected X-shaped array of ones, or it might consist of all configurations containing as many ones as zeros, or it might be any one of the other $2^{2^{ab}}$ possible subsets of C.) This very large set of possibilities constitutes the class of possible environments $\mathcal{E}$; it is the set of alternatives the adaptive plan must be prepared to handle. The adaptive device's task is to discover or "learn" which element of $\mathcal{E}$ is in force by learning what configurations belong to C_1. Then, when an arbitrary configuration is presented, the device can reliably indicate whether the configuration belongs to C_1, thereby detecting an instance of the pattern.

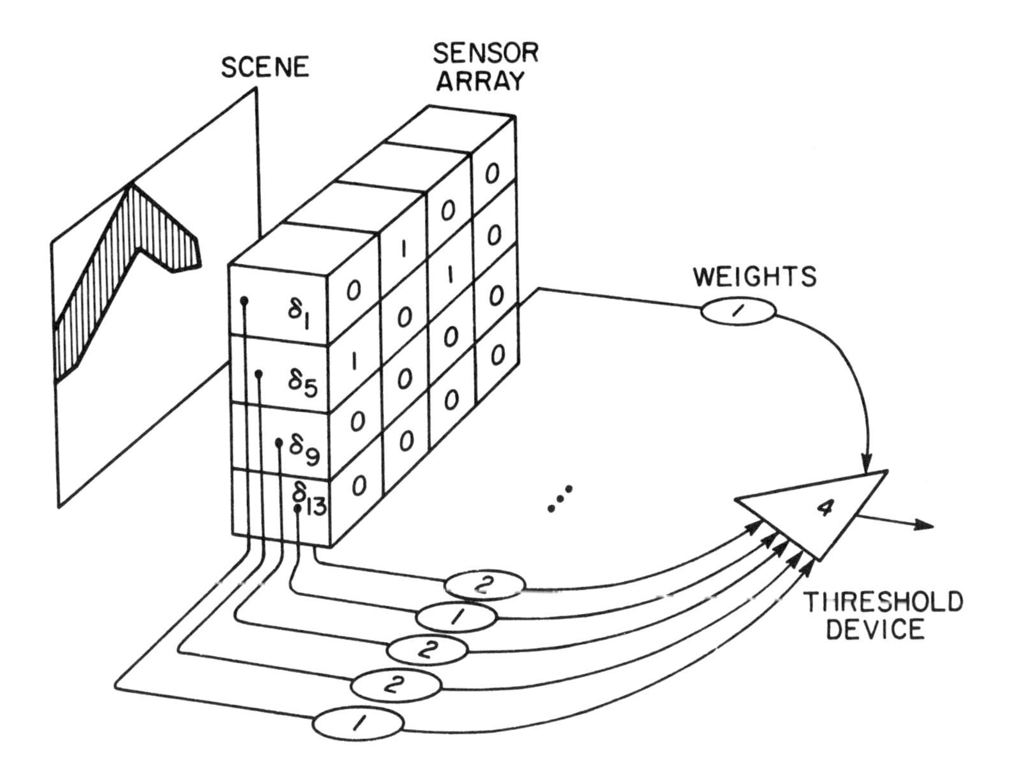

The scene shown is classified as C^+ because $\sum_{i=1}^{16} w_i\delta_i(t) = \delta_1(t) + 2\delta_2(t) + 2\delta_3(t) + \delta_4(t) + 2\delta_5(t) + 4\delta_6(t) + \cdots > 4$.

Fig. 1. A simple pattern recognizer

The particular pattern recognition device considered here—a linear threshold device—processes the input signals $\delta_i(t)$ by first multiplying each one by some weight w_i and then summing them to yield $\sum_{i=1}^{ab} w_i\delta_i(t)$. When this sum exceeds a given fixed threshold K the input configuration will be said to be a member of the set C^+, otherwise a member of the set C^-. (It should be clear that $C^+ \cup C^- = C$ and that $C^+ \cap C^-$ is empty, so that the linear threshold device partitions C into two classes.) More precisely C^+ is supposed to be an approximation to C_1, so that when the sum exceeds the fixed threshold K, the device indicates (rightly or wrongly) that the input configuration is an instance of the pattern. The object of the adaptive plan, then, is to discover as rapidly as possible a set of weights for which the partition (C^+, C^-) approximates the partition (C_1, C_0), so that $C^+ \cong C_1$ and $C^- \cong C_0$. (This device, as noted earlier, is quite limited; there are many partitions (C_1, C_0) that can only be poorly approximated by (C^+, C^-), no matter what set of weights is chosen.) Now, let $W = \{v_1, v_2, \ldots, v_k\}$ be the set of possible values for the weights w_i; that is, each $w_i \in W$, $i = 1, \ldots, ab$. Thus, with a fixed threshold K, the set of attainable structures $\mathcal{A}$ is the set of all *ab*-tuples, W^{ab}.

The natural performance measure, μ_E, relative to any particular partition $E \in \mathcal{E}$ is the proportion of all configurations correctly assigned (to C_1 and C_0). That is, μ_E maps each *ab*-tuple into the fraction of correct recognitions achieved thereby, a number in the interval [0, 1], $\mu_E: W^{ab} \rightarrow [0, 1]$. (In this example the outcome of *each* test—"configuration correctly classified" or "configuration incorrectly classified"—will be treated as the plan's input. The same *ab*-tuple may have to be tested repeatedly to establish an estimate of its performance.)

A simple plan τ_0 for discovering the best set of weights in W^{ab} is to try various *ab*-tuples, either in some predetermined order or at random, estimating the performance of each in its turn; the best *ab*-tuple encountered up to a given point in time is saved for comparison with later trials—this "best-to-date" *ab*-tuple being replaced immediately by any better *ab*-tuple encountered in a later trial. It should be clear that this procedure must eventually uncover the "best" *ab*-tuple in W^{ab}. But note that even for $k = 10$ and $a = b = 10$, W^{ab} has 10^{100} elements. This is a poor augury for any plan which must exhaustively search W^{ab}. And that is exactly what the plan just described must undertake, since the outcome of earlier tests in no way affects the ordering of later tests.

Let's look at a (fairly standard) plan τ_{00} which *does* use the outcome of each test to help determine the next structure for testing. The basic idea of this plan is to change the weights whenever a presentation is misassigned so as to decrease the likelihood of similar misassignments in the future. In detail: Let the values in W be ordered in increasing magnitude so that $v_{j+1} > v_j$, $j = 1, 2, \ldots, k - 1$ (for instance, the weights might be located at uniform intervals so that $v_{j+1} = v_j + \Delta$). Then the algorithm proceeds according to the following prescription:

1. If the presentation at time t is assigned to C_0 when it should have been assigned to C_1 then, for each i such that $\delta_i(t) = 1$, replace the corresponding weight by the next highest weight (in the case of uniform intervals the new weight would be the old weight w_i increased by Δ, $w_i + \Delta$). Leave the other weights unchanged.

2. If the presentation at time t is assigned to C_1 instead of C_0 then, for each i such that $\delta_i(t) = 1$, replace the corresponding weight by the next lowest weight (for uniform intervals, the new weight is $w_i - \Delta$).

We cannot yet fruitfully discuss the merits of this plan in comparison to alternatives; we can only note that the order in which τ_{00} tests the elements of $\mathcal{A}$ does indeed depend upon the information it receives. That is, the trajectory through $\mathcal{A} = W^{ab}$ is conditional on the outcomes $\mu_E(A)$, $A \in \mathcal{A}$, of prior tests.

4. A COMPLEX NATURAL ADAPTIVE SYSTEM

Here we will look at biological adaptation via changes in genetic makeup—the first of a series of progressively more detailed examinations. This section will present only biological facts directly relevant to adaptation, with a caveat to the reader about the dangers of unintentional emphasis and oversimplification inherent in such a partial picture.

It is a familiar fact (but one we will delve into later) that every organism is an amalgam of characteristics determined by the genes in its chromosomes. Each gene has several forms or alternatives—*alleles*—producing differences in the set of characteristics associated with that gene. (E.g., certain strains of garden pea have a single gene which determines blossom color, one allele causing the blossom to be white, the other pink; bread mold has a gene which in normal form causes synthesis of vitamin B_1, but several mutant alleles of the gene are deficient in this ability; human sickle cell anemia results from an abnormal allele of one of the genes determining the structure of hemoglobin—interestingly enough, in environments where malaria is endemic, the abnormal allele can confer an advantage.) There are tens of thousands of genes in the chromosomes of a typical vertebrate, each of which (on the evidence available) has several alleles. Taking the set of attainable structures $\mathcal{A}$ to be the set of chromosomes obtained by making all possible combinations of alleles, we see that $\mathcal{A}$ contains on the order of $2^{10,000} \cong 10^{3000}$ structures for a typical vertebrate species (assuming 2 alleles for each of 10,000 genes). Even a very large population, say 10 billion individuals of that species, contains only a minuscule fraction of the possibilities.

The enormous number of possible genetic structures—*genotypes*—for a single vertebrate species is an indicator of the complexity of such systems, but it is only an indicator. The basic complexity of these systems comes from the interactions of the genes. To see just how extensive these interactions are, it is worth looking briefly at some of the related biochemistry. Without going into detail, different alleles of the same gene produce related proteins, which in turn produce the variations in expressed characteristics associated with that gene. Typically these proteins (or combinations of them) are powerful biological catalysts called *enzymes*, capable of modifying reaction rates by factors of 10,000 and more. For this reason, genes exercise extensive control over the ongoing reactions in a cell—the enzymes they produce modulate ongoing reactions so strongly that they are the major determinants of the cell's form. Moreover, the products of any given enzyme-controlled reaction may, and generally do, enter into several subsequent reactions. Thus the effects of changes in a single enzyme are often widespread, causing gross changes in cell form and function. The human hereditary disorder called phenylketonuria results from an (undesirable) allele of a single gene; the presence of this allele has pronounced effects upon a whole battery of characteristics ranging from hair color and head size through intelligence. It is equally true that several genes may jointly determine a given characteristic, e.g., eye color in humans.

All of this adds considerably to the complexity of the system, but the greatest complexities come about because the effects of different enzymes are not additive—a phenomenon known as *epistasis.* For example, if a sequence of reactions depends upon several enzymes, for practical purposes the sequence does not proceed at all until all of the enzymes are present; subtraction of one enzyme stops the reaction completely. More complicated reactions involving positive and negative feedback are common, particularly those in which the output of a reaction sequence is a catalyst or inhibitor for some intermediate step of the reaction. The main point is that the effect of each allele depends strongly upon what other alleles are present and small changes can often produce large effects. The amalgam of observed characteristics—the *phenotype*—depends strongly upon these epistatic effects.

Because of epistasis there is no simple way to apportion credit to individual alleles for the performance of the resulting phenotype. What may be a good allele when coordinated with an appropriate set of alleles for other genes, can be disastrous in a different genetic context. Thus adaptation cannot be accomplished by selecting among the alleles for one gene independently of what alleles appear for other genes. The problem is like the problem of adjusting the "height," "vertical linearity," and "vertical hold" controls on a television set. A "best setting" for "height," ignoring the settings of the other two controls, will be destroyed as soon as one attempts to better the setting of either of the other two controls. The problem is vexing enough when there are three interdependent controls, as anyone who has attempted these adjustments can testify, but it pales in comparison to the genetic case where dozens or hundreds of interdependent alleles can be involved. Roughly, the difficulty of the problem increases by an order of magnitude for each additional gene when the interdependencies are intricate (but see the discussions in chapter 4 and pp. 160–61).

Given the pervasiveness of epistasis, adaptation via changes in genetic makeup becomes primarily a search for *coadapted* sets of alleles—alleles of different genes which together significantly augment the performance of the corresponding phenotype. (In chapter 4 the concept of a coadapted set of alleles will be generalized, under the term *schema*, to the point where it applies to the full range of adaptive systems.) It should be clear that coadaptation depends strongly upon the environment of the phenotype. The large coadapted set of alleles which produces gills in fish augments performance only in aquatic environments. This dependence of coadaptation upon characteristics of the environment gives rise to the notion of an *environmental niche*, taken here to mean a set of features of the environment which can be exploited by an appropriate organization of the phenotype. (This is a broader interpretation than the usual one which limits niche to those environmental features particularly exploited by a given species.) Examples of environmental niches fitting this interpretation are: (i) an oxygen-poor, sulfur-rich environment such as is found at the bottom of ponds with large amounts of decaying matter—a class of anaerobic bacteria, the thiobacilli, exploits this niche by means of a complex of enzymes enabling them to use sulfur in place of oxygen to carry out oxidation; (ii) the "bee-rich" environment exploited by the orchid *Ophrys apifera* which has a flower mimicking the bee closely enough to induce pollination via attempted copulation by the male bees; (iii) the environment rich in atmospheric vibrations in the frequency range of 50 to 50,000 cycles per second—the bones of the mammalian ear are a particular adaptation of parts of the reptilian jaw which aids in the detection of these vibrations, an adaptation which clearly must be coordinated with many other adaptations, including a sophisticated information-processing network, before it can improve an organism's chances of survival. It is important to note that quite distinct coadapted sets of alleles can exploit the same environmental niche. Thus, the eye of aquatic mammals and the (functionally similar) eye of the octopus exploit the same environmental niche, but are due to coadapted sets of alleles of entirely unrelated sets of genes.

The various environmental niches $E \in \mathcal{E}$ define different opportunities for adaptation open to the genetic system. To exploit these opportunities the genetic system must select and use the sets of coadapted alleles which produce the appropriate phenotypic characteristics. The central question for genetic systems is: How are initially unsuited structures transformed to (an observed range of) structures suited to a variety of environmental niches $\mathcal{E}$? To attempt a general answer to this question we need a well-developed formal framework. The framework available at this point is insufficient, even for a careful description of a candidate adaptive plan τ for genetic systems, unlike the case of the simpler artificial system. A fortiori, questions about such adaptive plans, and critical questions about efficiency, must wait upon further development of the framework. We *can* explore here some of the requirements an adaptive plan τ must meet if it is to be relevant to data about genetics and evolution.

In beginning this exploration we can make good use of a concept from mathematical genetics. The action of the environment $E \in \mathcal{E}$ upon the phenotype (and thereby upon the genotype $A \in \mathcal{A}$) is typically summarized in mathematical studies of genetics by a single performance measure μ_E called *fitness.* Roughly, the fitness of a phenotype is the number of its offspring which survive to reproduce (precise definitions will be given later in connection with the appropriate formal models, see section 3.1). This measure rests upon a universal, and familiar, feature of biological systems: Every individual (phenotype) exists as a member of a population of similar individuals, a population constantly in flux because of the reproduction and death of the individuals comprising it. The fitness of an individual is clearly related to its influence upon the future development of the population. When many offspring of a given individual survive to reproduce, then many members of the resulting population, the "next generation," will carry the alleles of that individual. Genotypes and phenotypes of the next generation will be influenced accordingly.

Fitness, viewed as a measure of the genotype's influence upon the future, introduces a concept useful through the whole spectrum of adaptation. A good way to see this concept in wider context is to view the testing of genotypes as a sampling procedure. The sample space in this case is the set of all genotypes $\mathcal{A}$ and the outcome of each sample is the performance μ_E of the corresponding phenotype. The general question associated with fitness, then, is: To what extent does the outcome $\mu_E(A)$ of a sample $A \in \mathcal{A}$ influence or alter the sampling plan τ (the kinds of samples to be taken in the future)? Looking backward instead of forward, we encounter a closely related question: How does the history of the outcomes of previous samples influence the current sampling plan? The answers to these questions go far toward determining the basic character of any adaptive process.

We have already seen that the answer to the first question, for genetic systems, is that the future influence of each individual $A \in \mathcal{A}$ is directly proportional to the sampled performance $\mu_E(A)$. This relation need not be so in general—

there are many well-established procedures for optimization, inference, mathematical learning, etc., where the relation between sampled performance and future sampling is quite different. Nevertheless reproduction in proportion to measured performance is an important concept which can be generalized to yield sampling plans—*reproductive plans*—applicable to any adaptive problem (including the broad class of problems where there is no natural notion of reproduction). Moreover, once reproductive plans have been defined in the formal framework, it can be proved that they are efficient (in a reasonable sense) over a very broad range of conditions.

A part of the answer to the second question, for genetic systems, comes from the observation that future populations can only develop via reproduction of individuals in the current population. Whatever history is retained must be represented in the current population. In particular, the population must serve as a summary of observed sample values (performances). The population thereby has the same relation to an adaptive process that the notion of (complete) state has to the laws of physics or the transition functions of automata theory. Knowing the population structure or state enables one to determine the future without any additional information about the past of the system. (That is, different sampling sequences which arrive at the same population will have exactly the same influence on the future.) The state concept has been used as a foundation stone for formal models in a wide variety of fields; in the formal development to follow generalizations of population structure will have this role.

An understanding of the two questions just posed leads to a deeper understanding of the requirements on a genetic adaptive plan. It also leads to an apparent dilemma. On the one hand, if offspring are simple duplicates of fit members of the population, fitness is preserved but there is no provision for improvement. On the other hand, letting offspring be produced by simple random variation (a process practically identical to enumeration) yields a maximum of new variants but makes no provision for retention of advances already made. The dilemma is sharpened by two biological facts: (1) In biological populations consisting of advanced organisms (say vertebrates) no two individuals possess identical chromosomes (barring identical twins and the like). This is so even if we look over many (all) successive generations. (2) In realistic cases, the overwhelming proportion of *possible* variants (all possible allele combinations, not just those observed) are incapable of surviving to produce offspring in the environments encountered. Thus, by observation (1), advances in fitness are not retained by simple duplication. At the same time, by observation (2), the observed lack of identity cannot result from simple random variation because extinction would almost certainly follow in a single generation—variants chosen completely at random are almost certain to be sterile.

In attempting to see how this "dilemma" is resolved, we begin to encounter some of the deeper questions about adaptation. We can only hint at the dilemma's resolution in this preliminary survey. Even a clear statement of the resolution requires a considerable formal structure, and proof that it is in fact a resolution requires still more effort. Much of the understanding hinges on posing and answering two questions closely related to the questions generated by the concept of fitness: How can an adaptive plan τ (specifically, here a plan for genetic systems) retain useful portions of its (rapidly growing) history along with advances already made? How is the adaptive plan τ to access and use its history (the portion stored) to increase the likelihood of fit variants ($A \in \mathcal{A}$ such that $\mu_E(A)$ is above average)? Once again these are questions relevant to the whole spectrum of fields mentioned at the outset.

The resolution of the dilemma lies in the action of the genetic operators Ω within the reproductive plan τ. The best-known genetic operators exhibit two properties strongly affecting this action: (1) The operators do not directly affect the size of the population—their main effect is to alter and redistribute alleles within the population. (The alleles in an individual typically come from more than one source in the previous generation, the result, for example, of the mating of parents in the case of vertebrates, or of transduction in the case of bacteria.) (2) The operators infrequently separate alleles which are close together on a chromosome. That is, alleles close together typically remain close together after the operators have acted.

Useful clues to the dilemma's resolution emerge when we look at the effect of these operators in a simple reproductive plan, τ_1. This plan can be thought of as unfolding through repeated application of a two-phase procedure: During phase one, additional copies of (some) individuals exhibiting above-average performance are added to the population while (some) individuals of subaverage performance are deleted. More carefully, each individual has an expected number of offspring, or rate of reproduction, proportional to its performance. (If the population is to be constant in size, the rates of reproduction must be "normalized" so that their average over the population at any time is 1.) During phase two, the genetic operators in Ω are applied, interchanging and modifying sets of alleles in the chromosomes of different individuals, so that the offspring are no longer identical to their progenitors. The result is a new, modified population. The process is iterated to produce successive generations of variants.

More formally, in an environment which assigns an observable performance to each individual, τ_1 acts as follows: At the beginning of each time period t, the plan's accumulated information about the environment resides in a finite population $\mathcal{A}(t)$ selected from $\mathcal{A}$. The most important part of this information is given by the discrete distributions which give the proportions of different sets of alleles in the population $\mathcal{A}(t)$. $\mathcal{A}(t)$ serves not only as the plan's repository of accumulated information, but also as the source of new variants which will give rise to $\mathcal{A}(t+1)$. As indicated earlier, the formation of $\mathcal{A}(t+1)$ proceeds in two phases. During the first phase, $\mathcal{A}(t)$ is modified to form $\mathcal{A}'(t)$ by copying each individual in $\mathcal{A}(t)$ a number of times dependent upon the individual's observed performance. The number of copies made will be determined stochastically so that the expected number of copies increases in proportion to observed performance. During the second phase, the operators are applied to the population $\mathcal{A}'(t)$, interchanging and modifying the sets of alleles, to produce the new generation $\mathcal{A}(t+1)$.

One key to understanding τ_1's resolution of the dilemma lies in observing what happens to small sets of adjacent alleles under its action. In particular, what happens if an adjacent set of alleles appears in several different chromosomes of above-average fitness and not elsewhere? Because each of the chromosomes will be duplicated an above-average number of times, the given alleles will occupy an increased proportion of the population after the duplication phase. This increased proportion will of course result whether or not the alleles had anything to do with the above-average fitness. The appearance of the alleles in the extra-fit chromosomes might be happenstance, but it is equally true that any correlation between the given selection of alleles and above-average fitness will be exploited by this action. Moreover, the more varied the chromosomes containing the alleles, the less likely it is that the alleles and above-average fitness are uncorrelated.

What happens now when the genetic operators Ω are applied to form the next generation? As indicated earlier, the closer alleles are to one another in the chromosome the less likely they are to be separated during the operator phase. Thus the operator phase usually transfers adjacent sets of genes as unit, placing them in new chromosomal contexts without disturbing them otherwise. These new contexts further test the sets of alleles for correlation with above-average fitness. If the selected set of alleles does indeed augment fitness, the chromosomes containing the set will again (on the average) be extra fit. On the other hand, if the prior associations were simply happenstance, sustained association with extra-fit chromosomes becomes increasingly less likely as the number of trials (new contexts) increases. The net effect of the genetic plan over several generations will be an increasing predominance of alleles and sets of alleles augmenting fitness in the given environment.

In observing what happens to small sets of genes under its action, we have seen one way in which the plan τ_1 preserves the history of its interactions with the environment. It also retains certain kinds of advances thereby, favoring structural components which have proved their worth by augmenting fitness. At the same time, since these components are continually tried in new contexts and combinations, stagnation is avoided. In brief, sets of alleles engendering above-average performance provide comparative success in reproduction for the chromosomes carrying them. This in turn assures that these alleles become predominant components of later generations of chromosomes. Though this description is sketchy, it does indicate that reproductive plans using genetic operators proceed in a way which is neither enumeration nor simple duplication of fit structures. The full story is both more intricate and more sophisticated. Because reproductive plans are provably efficient over a broad range of conditions, we will spend considerable time later unraveling the skeins of this story.

5. SOME GENERAL OBSERVATIONS

One point which comes through clearly from the examples is the enormous size of $\mathcal{A}$, even for a very modest system. This size has a fatal bearing on what is at first sight a candidate for a "universal" adaptive plan. The candidate, called τ_0 in the first example, and henceforth designated an *enumerative* plan, exhaustively tests the structures in $\mathcal{A}$. Enumerative plans are characterized by the fact that the order in which they test structures is unaffected by the outcome of previous tests. For example, the plan first generates and tests all structures attainable (from an initially given structure) by single applications of the basic operators, then all structures attainable by two applications of the operators, etc. The plan preserves the fittest structure it has encountered up to any given point in the process, replacing that structure immediately upon generating a structure which is still more fit. Thus, given enough time (and enough stability of the environment so that the fitness of structures does not change during the process) an enumerative plan is guaranteed to discover the structure most fit for any environment confronting it. The simplicity of this plan, together with the guarantee of discovering the most fit structure, would seem to make it a very important adaptive plan. Indeed enumerative plans have been repeatedly proposed and studied in most of the areas mentioned in section 1.1. They are often set forth in a form not obviously enumerative, particularly in evolutionary studies (mutation in the absence of other genetic operators), learning (simple trial-and-error), and artificial intelligence (random search).

However, in all but the most constrained situations, enumerative plans are a false lead.

The flaw, and it is a fatal one, asserts itself when we begin to ask, "How long is eventually?" To get some feeling for the answer we need only look back at the first example. For that very restricted system there were 10^{100} structures in $\mathcal{A}$. In most cases of real interest, the number of possible structures vastly exceeds this number, and for natural systems like the genetic systems we have already seen that numbers like $2^{10,000} \cong 10^{3000}$ arise. If 10^{12} structures could be tried every second (the fastest computers proposed to date could not even add at this rate), it would take a year to test about $3 \cdot 10^{19}$ structures, or a time vastly exceeding the estimated age of the universe to test 10^{100} structures.

It is clear that an attempt to adapt by means of an enumerative plan is foredoomed in all but the simplest cases because of the enormous times involved. This extreme inefficiency makes enumerative plans uninteresting either as hypotheses about natural processes or as algorithms for artificial systems. It follows at once that an adaptive plan cannot be considered good simply because it will eventually produce fit structures for the environments confronting it; it must do so in a reasonable time span. What a "reasonable time span" is depends strongly on the environments (problems) under consideration, but in no case will it be a time large with respect to the age of the universe. This question of efficiency or "reasonable time span" is the pivotal point of the most serious contemporary challenge to evolutionary theory: Are the known genetic operators sufficient to account for the changes observed in the alloted geological intervals? There is of course evidence for the existence of adaptive plans much more efficient than enumeration. Arthur Samuel (1959) has written a computer program which learned to play tournament calibre checkers, and humans do manage to adapt to very complex environments in times considerably less than a century. It follows that a major part of any study of the adaptive process must be the discovery of factors which provide efficiency while retaining the "universality" (robustness) of enumeration. It does not take analysis to see that an enumerative plan is inefficient just because it always generates structures in the same order, regardless of the outcome of tests on those structures. The way to improvement lies in avoiding this constraint.

The foregoing points up again the critical nature of the adaptive plan's initial uncertainty about its environment, and the central role of the procedures it uses to store and access the history of its interactions with that environment. Since different structures perform differently in different environments, the plan's task is set by the aspects of the environment which are unknown to it initially. It must generate structures which perform well (are fit) in the particular environment confronting it, and it must do this efficiently. Interest centers on *robust* adaptive plans—plans which are efficient over the range of environments $\mathcal{E}$ they may encounter. Giving robustness precise definition and discovering something of the factors which make an adaptive plan robust is the formal distillation of questions about efficiency. Because efficiency is critical, the study of robustness has a central place in the formal development.

The discussion of genetic systems emphasized two general requirements bearing directly on robustness: (1) The adaptive plan must retain advances already made, along with portions of the history of previous plan-environment interactions. (2) The plan must use the retained history to increase the proportion of fit structures generated as the overall history lengthens. The same discussion also indicated the potential of a particular class of adaptive plans—the reproductive plans. One of the first tasks, after setting out the formal framework, will be to provide a general definition of this class of plans. Lifting the reproductive plans from the specific genetic context makes them useful across the full spectrum of fields in which adaptation has a role. This widened role for reproductive plans can be looked upon as a first validation of the formalism. A much more substantial validation follows closely upon the definition, when the general robustness of reproductive plans is proved via the formalism. Later we will see how reproductive plans using generalized genetic operators retain and exploit their histories. Throughout the development, reproductive plans using genetic operators will serve to illuminate key features of adaptation and, in the process, we will learn more of the robustness, wide applicability, and general sophistication of such plans.

Summarizing: This entire survey has been organized around the concept of an adaptive plan. The adaptive plan, progressively modifying structure by means of suitable operators, determines what structures are produced in response to the environment. The set of operators Ω and the domain of action of the adaptive plan $\mathcal{A}$ (i.e., the attainable structures) determine the plan's options; the plan's objective is to produce structures which perform well in the environment E confronting it. The plan's initial uncertainty about the environment—its room for improvement—is reflected in the range of environments $\mathcal{E}$ in which it may have to act. The related performance measures μ_E, $E \in \mathcal{E}$, change from environment to environment since the same structure performs differently in different environments. These objects lie at the center of the formal framework set out in chapter 2. Chapter 3 provides illustrations of the framework as applied to genetics, economics, game-playing, searches, pattern recognition, statistical inference, control, function optimization, and the central nervous system.

A brief look at the enormous times taken by enumerative plans to discover fit structures, even when the domain of action $\mathcal{A}$ is greatly constrained, makes it clear that efficiency is a *sine qua non* of studies of adaptation. Efficiency acts as a cutting edge, shearing away plans "too slow" to serve as hypotheses about natural systems or as algorithms for artificial systems. Whether an adaptive plan is to serve as hypothesis or algorithm, information about its robustness—its efficiency in the environments $\mathcal{E}$—is critical. The latter part of this book will be much concerned with this topic. Chapter 4 introduces a critical tool for the investigation and construction of efficient adaptive plans—schemata. This generalization of coadapted sets of alleles provides an efficient way of defining and exploiting properties associated with above-average performance. Chapter 5 develops a criterion for measuring the efficiency with which adaptive plans *improve* average performance and then relates this criterion to the exploitation of schemata. Chapter 6 introduces generalized genetic plans and chapter 7 establishes their robustness. Chapter 8 studies mechanisms which enable genetic plans to use predictive modeling for flexible exploitation of the large fluxes of information provided by typical environments.

The emphasis throughout the book is on general principles which help to resolve the problems and questions raised in this chapter. One particular interest will be the solution of problems involving hundreds to hundreds of thousands of interdependent parameters and multitudes of local optima—problems which largely lie outside the prescriptions of present day computational mathematics.

2. A Formal Framework

1. DISCUSSION

Three associated objects occupied the center of the preliminary survey.

> E, the environment of the system undergoing adaptation,
> τ, the adaptive plan which determines successive structural modifications in response to the environment,
> μ, a measure of the performance of different structures in the environment.

Implicit in the discussion is a decomposition of the overall process into two disjoint parts—the adaptive system employing τ, and its environment E. This decomposition is usually fixed or strongly suggested by the particular emphasis of each study, but occasionally it can be arbitrary and, rarely, it can be a source of difficulty. Thus, in some biological studies the epidermis naturally serves as the adaptive system-environment boundary, while in other biological studies we deal with populations which have no fixed spatial boundaries, and in ecological settings the boundary shifts with every change in emphasis. Similarly, the emphasis of the study usually determines what notion of performance is relevant and how it is to be measured to yield μ. Because E, τ, and μ are central and can be regularly identified in problems of adaptation, the formal framework will be built around them.

In the basic formalism the adaptive plan τ will be taken to act at discrete instants of time, $t = 1, 2, 3, \ldots$, rather than continuously. The primary reason for adopting a discrete time-scale is the simpler form it confers on most of the important results. Also this formalism intersects smoothly with extant mathematical theories in several fields of interest where much of the development is based on a discrete time-scale, viz., mathematical economics, sequential sampling theory, the theory of self-reproducing automata, and major portions of population genetics. Where continuity is more appropriate, it is often straightforward to obtain continuous counterparts of definitions and theorems, though in some cases appropriate restatements are full-fledged research problems with the discrete results serving only as guidelines. In any case, the instants of time can be freely reinterpreted in different applications—they may be nanoseconds in one application (e.g., artificial intelligence), centuries in another (e.g., evolutionary theory). The properties and relations established with the formalism remain valid, only their durations will vary. Thus, at the outset, we come upon a major advantage of the formalism: Features or procedures easily observed in one process can be abstracted, set within the framework, and analyzed so that they can be interpreted in other processes where duration of occurrence, or other detail, obscures their role.

As our starting point for constructing the formalism let us take the domain of action of the adaptive plan, the set of *structures* $\mathcal{A}$. At the most abstract level $\mathcal{A}$ will simply be an arbitrary, nonempty set; when the theory is applied, $\mathcal{A}$ will designate the set of structures appropriate to the field of interest. Because the more general parts of the theory are valid for any nonempty set $\mathcal{A}$, we have great latitude in interpreting or applying the notion of structure in particular cases. Stated the other way around, the diversity of objects which can serve as elements of $\mathcal{A}$ assures flexibility in applying the theory. In practice, the elements of $\mathcal{A}$ can be the formal counterparts of objects much more complex than the basic structures (chromosomes, mixes of goods, etc.) of the preliminary survey. They may be sets, sequences, or probability distributions over the basic structures; moreover, portions of the adaptive system's past history may be explicitly represented as part of the structure. Often the basic structures themselves will exhibit additional properties, being presented as compositions of interacting components (chromosomes composed of alleles, programs composed of sets of instructions, etc.). Thus (referring to section 1.4), if the elements of $\mathcal{A}$ are to represent chromosomes with ℓ specified genes, where the ith gene has a set of k_i alleles $A_i = \{a_{i1}, \ldots, a_{ik_i}\}$, then the set of structures becomes the set of all combinations of alleles,

$$\mathcal{A} = A_1 \times A_2 \times \cdots \times A_\ell = \Pi_{i=1}^{\ell} A_i.$$

Finally, the set $\mathcal{A}$ will usually be potential rather than actual. That is, elements become available to the plan only by successive modification (e.g., by rearrangement of components or construction from primitive elements), rather than by selection from an extant set. We will examine all of these possibilities as we go along, noting that relevant elaborations of the elements of $\mathcal{A}$ provide a way of specializing the general parts of the theory for particular applications.

The *adaptive plan* τ produces a sequence of structures, i.e., a trajectory through $\mathcal{A}$, by making successive selections from a set of *operators* Ω. The particular selections made are influenced by information obtained from the environment E, so that the plan τ typically generates different trajectories in different environments. The adaptive system's ability to discriminate among various environments is limited by the range I of stimuli or *signals* it can receive. More formally: Let the structure tried at time t be $\mathcal{A}(t) \in \mathcal{A}$. Then the particular environment E confronting the adaptive system reacts by producing a signal $I(t)$. Different structures may of course be capable of receiving different ranges of signals. That is, if I_A is the range of signals which A can receive, then for $A' \neq A$ it may be that $I_A \neq I_{A'}$. To keep the presentation simple, I is used to designate the *total* range of signals $\bigcup_{A \in \mathcal{A}} I_A$ receivable by structures in $\mathcal{A}$. The particular information $I(t)$ received by the adaptive system at time t will then be constrained to the subset of signals $I_{\mathcal{A}(t)} \subset I$ which the structure at time t, $\mathcal{A}(t)$, can receive. I may have many components corresponding, say, to different sensors. Thus, referring to the example of section 1.3, I consists of ab components $I_1 \times I_2 \times \cdots \times I_{ab} = \Pi_{i=1}^{ab} I_i$. In this case $I_i = \{0,1\}$ for all i since the ith component of I represents the range of values the ith sensor δ_i

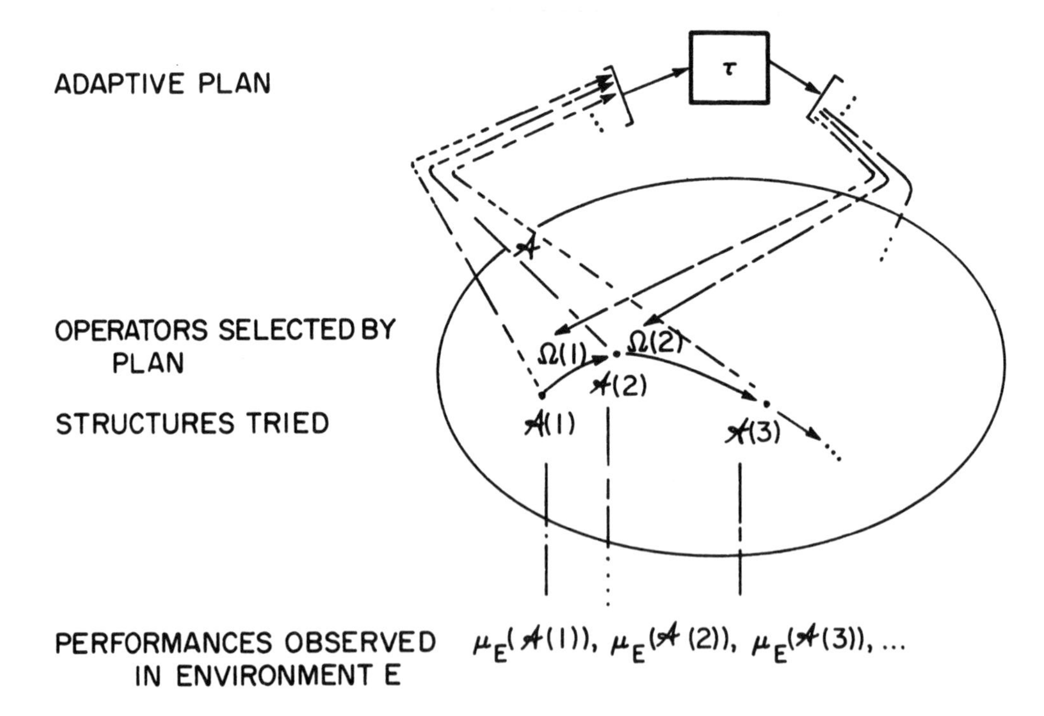

Fig. 2. Schematic of adaptive plan's operation

can transmit. That is, given a particular signal $I(t) \in I$ at time t, the ith component $I_i(t)$ is the value $\delta_i(t)$ of the ith sensor at time t. In general the sets I_i may be quite different, corresponding to different kinds of sensors or sensory modalities.

The formal presentation of an adaptive plan τ can be simplified by requiring that $\mathcal{A}(t)$ serve as the state of the plan at time t. That is, in addition to being the structure tried at time t, $\mathcal{A}(t)$ must summarize whatever accumulated information is to be available to τ. We have just provided that the total information received by τ up to time t is given by the sequence $\langle I(1), I(2), \ldots, I(t-1)\rangle$. Generally only part of this information is retained. To provide for the representation of the retained information we can make use of the latitude in specifying $\mathcal{A}$. Think of $\mathcal{A}$ as consisting of two components $\mathcal{A}_1$ and $\mathcal{M}$, where $\mathcal{A}_1(t)$ is the structure tested against the environment at time t, and the *memory* $\mathcal{M}(t)$ represents other retained parts of the *input history* $\langle I(1), I(2), \ldots, I(t-1)\rangle$. Then the plan can be represented by the two-argument function

$$\tau: I \times \mathcal{A} \rightarrow \mathcal{A}.$$

Here the structure to be tried at time $t+1$, $\mathcal{A}_1(t+1)$, along with the updated memory $\mathcal{M}(t+1)$, is given by

$$(\mathcal{A}_1(t+1), \mathcal{M}(t+1)) = \mathcal{A}(t+1) = \tau(I(t), \mathcal{A}(t)) = \tau(I(t), (\mathcal{A}_1(t), \mathcal{M}(t))).$$

(The projection of τ on $\mathcal{M}$,

$$\tau_{\mathcal{M}}: I \times \mathcal{A}_1 \times \mathcal{M} \rightarrow \mathcal{M}$$

defined so that

$$\tau_{\mathcal{M}}(I(t), \mathcal{A}_1(t), \mathcal{M}(t)) = \text{proj}_2\,[\tau(I(t), \mathcal{A}(t))] = \mathcal{M}(t+1)$$

is that part of τ which determines how the plan's memory is updated.) It is clear that any theorems or interpretations established for the simple form

$$\tau: I \times \mathcal{A} \rightarrow \mathcal{A}$$

can at once be elaborated, without loss of generality or range of application, to the form

$$\tau: I \times (\mathcal{A}_1 \times \mathcal{M}) \rightarrow (\mathcal{A}_1 \times \mathcal{M}).$$

Thus the framework can be developed in terms of the simple, two-argument form of τ, elaborating it whenever we wish to study the mechanisms of trial selection or memory update in greater detail.

In what follows it will often be convenient to treat the adaptive plan τ as a stochastic process; instead of determining a unique structure $\mathcal{A}(t+1)$ from $I(t)$ and $\mathcal{A}(t)$, τ assigns probabilities to a range of structures and then selects accordingly. That is, given $I(t)$, $\mathcal{A}(t)$ may be transformed into any one of several structures $A'_1, A'_2, \ldots, A'_j, \ldots$, the structure A'_j being selected with probability P'_j. More formally: Let $\mathcal{P}$ be a set of admissible probability distributions over $\mathcal{A}$. Then

$$\tau: I \times \mathcal{A} \rightarrow \mathcal{P}$$

will be interpreted as assigning to each pair $(I(t), \mathcal{A}(t))$ a particular distribution over $\mathcal{A}$, $\mathcal{P}(t+1) \in \mathcal{P}$. The structure $\mathcal{A}(t+1)$ to be tried at time $t+1$ will then be determined by drawing a random sample from $\mathcal{A}$ according to the probability distribution $\mathcal{P}(t+1) = \tau(I(t), \mathcal{A}(t))$. For those cases where the plan τ is to determine the next structure $\mathcal{A}(t+1)$ uniquely, the distribution $\mathcal{P}(t+1)$ simply becomes a degenerate, one-point distribution where a single structure in $\mathcal{A}$ is assigned probability 1. Hence the form

$$\tau: I \times \mathcal{A} \rightarrow \mathcal{P}$$

includes the previous

$$\tau: I \times \mathcal{A} \rightarrow \mathcal{A}$$

as a special case.

In practice the transformation of $\mathcal{A}(t)$ to $\mathcal{A}(t+1)$ is usually accomplished by the application of an operator from some specified set of operators Ω. Thus the detailed operation of the adaptive plan τ is given by a function

$$\tau': I \times \mathcal{A} \rightarrow \Omega$$

and the set of operators

$$\Omega = \{\omega: \mathcal{A} \rightarrow \mathcal{P}\}$$

where the stochastic aspect is now embodied in the operators. If

$$\omega_t = \tau'(I(t), \mathcal{A}(t))$$

designates the particular operator selected by τ' at time t, then

$$\mathcal{P}(t+1) = \omega_t(\mathcal{A}(t)) = [\tau'(I(t), \mathcal{A}(t))](\mathcal{A}(t))$$

gives the resulting distribution over $\mathcal{A}$. Hence τ' determines τ once the functions in Ω are specified:

$$\tau(I(t), \mathcal{A}(t)) = [\tau'(I(t), \mathcal{A}(t))](\mathcal{A}(t)) = \mathcal{P}(t+1).$$

That is, the range of τ' can be changed from Ω to $\mathcal{P}$ with τ' being redefined so that

$$\tau'(I(t), \mathcal{A}(t)) = [\tau'(i(t), \mathcal{A}(t))](\mathcal{A}(t)) = \mathcal{P}(t+1).$$

With this extension τ' and τ become identical; for this reason one symbol "τ" will be used to designate both functions, the range being specified whenever the distinction is important.

The general objective of this formalism is comparison of adaptive plans, either as hypotheses about natural phenomena or as algorithms for artificial systems. The comparison naturally centers on the efficiency of different plans in locating high performance structures under a variety of environmental conditions. For a comparison to be made there must be a set of plans, given either explicitly or implicitly, which are candidates for comparison. This set will be formally designated $\mathfrak{J}$. Often $\mathfrak{J}$ will be the set of all possible plans employing the operators in Ω, but in some cases there will be constraints restricting $\mathfrak{J}$, while in others $\mathfrak{J}$ will be enlarged to include all possible plans over $\mathcal{A}$ (i.e., all possible functions of the form $\tau: I \times \mathcal{A} \to \mathcal{P}$). $\mathfrak{J}$, however defined, represents the set of technical or feasible options for the adaptive system under consideration.

As indicated in the survey, a nontrivial problem of adaptation exists only when the adaptive plan is faced with an initial uncertainty about its environment. This uncertainty is formalized by designating the set $\mathcal{E}$ of alternatives corresponding to characteristics of the environment unknown to the adaptive plan. The dependence of the plan's action upon the environment finds its formal counterpart in the dependence of the input $I(t)$ upon which *environment* $E \in \mathcal{E}$ actually confronts the plan. One case of particular importance is that in which the adaptive plan receives a direct indication of the performance of each structure it tries. That is, a part of the input $I(t)$ will be the *payoff* $\mu_E(\mathcal{A}(t))$ determined by the function

$$\mu_E : \mathcal{A} \to \text{Reals}$$

which measures the performance of each structure in the given environment.

Sometimes, when the performance of a structure in the environment E depends upon random factors, it is useful to treat the utility function as assigning a random variable from some predetermined set $\mathcal{U}$ to each structure in $\mathcal{A}$. Thus

$$\mu_E : \mathcal{A} \to \mathcal{U}$$

and the payoff assigned to $\mathcal{A}(t)$ is determined by a trial of the random variable $\mu_E(\mathcal{A}(t)) = \mathcal{U}(t)$. This extension does not add any generality to the framework (and hence is unnecessary at the abstract level) because any randomness involved in the interaction between the adaptive system and its environment can be subsumed in the stochastic action of the operators. (See chapter 5 and section 7.2, however.)

Much can be learned about adaptive plans in general by studying plans which act *only* in terms of payoff, plans for which

$$I(t) = \mu_E(\mathcal{A}(t)).$$

In particular, plans which receive information in addition to payoff should do at least as well as plans which receive only payoff information. Thus, the efficiency of payoff-only plans with respect to $\mathcal{E}$ sets a nontrivial lower bound on the efficiency of other plans.

To pose a problem in adaptation unambiguously one more element is required: a *criterion* χ for comparing the efficiency of different plans $\tau \in \mathcal{J}$ under the uncertainty represented by $\mathcal{E}$. Such a criterion must of necessity be fairly sophisticated since it must somehow take into account the varying efficiency of a plan in different environments. Thus, even with a definite measure of efficiency such as the average rate of increase of payoff, there is still the problem of variations across the environments $\mathcal{E}$. How is a plan which is highly efficient only in some subset of $\mathcal{E}$ to be compared with a plan which is moderately efficient in all the environments in $\mathcal{E}$? It should be clear that the plan favored will often depend upon the particular application. In spite of this there are some broadly based criteria which have quite general applicability. The simplest of these requires that a plan accumulate payoff in each $E \in \mathcal{E}$ more rapidly than an enumerative plan which has the same domain of action $\mathcal{A}$. The intuitive content of this criterion is clear: A plan which does not accumulate payoff at least as rapidly as the extremely inefficient enumerative plans should, except in simple situations, be eliminated as a hypothesis (about natural systems) or an algorithm (for artificial systems). Because it is often useful to smooth out short-term variations in judging a plan, several broadly based criteria are stated in terms of the long-term average rate of payoff. When the adaptive plan has the deterministic form $\tau: I \times \mathcal{A} \rightarrow \mathcal{A}$, other, more general criteria are based on the cumulative payoff function

$$U_{\tau,E}(T) = \sum_{t=1}^{T} \mu_E(\mathcal{A}(\tau, t))$$

where $\mathcal{A}(\tau, t)$ is the structure selected by τ in E at time t, $\mu_E(\mathcal{A}(\tau, t))$ is the corresponding payoff, and $U_{\tau,E}(T)$ is the total payoff received by τ in E to time T. (The average rate of payoff is just the function $U_{\tau,E}(T)/T$ based on the cumulative payoff function $U_{\tau,E}(T)$.) When the adaptive plan is stochastic, $\tau: I \times \mathcal{A} \rightarrow \mathcal{P}$, it is natural to substitute the expected payoff under $\mathcal{P}(t)$, $\bar{\mu}_E(\tau, t)$, for $\mu_E(\mathcal{A}(\tau, t))$. (If $\mathcal{A}$ is countable, $\bar{\mu}_E(\tau, t)$ is simply given by $\bar{\mu}_E(\tau, t) = \sum_j \mathcal{P}(A_j, t)\mu_E(A_j)$ where $\mathcal{P}(A_j, t)$ is the probability of selecting $A_j \in \mathcal{A}$ when the distribution over $\mathcal{A}$ is $\mathcal{P}(t)$.) Thus, for stochastic adaptive plans,

$$U_{\tau,E}(T) = \sum_{t=1}^{T} \bar{\mu}_E(\tau, t).$$

Following this line, a useful performance target can be formulated in terms of the greatest possible cumulative payoff in the first T time-steps,

$$U_E^*(T) = \operatorname*{lub}_{\tau \in \mathfrak{J}} U_{\tau,E}(T).$$

An important criterion, appearing frequently in the literature of control theory and mathematical economics (see chapter 3, "Illustrations"), can be concisely formulated in terms of U_E^*: τ accumulates payoff at an *asymptotic optimal rate* if

$$\lim_{T\to\infty} [(U_{\tau,E}(T)/T)/(U_E^*(T)/T)] = \lim_{T\to\infty} [U_{\tau,E}(T)/U_E^*(T)] = 1.$$

In other words, the rate at which τ accumulates payoff is, in the limit, the same as the best possible rate. Often it is desirable to have a much stronger criterion setting standards on *interim* behavior. That is, even though the payoff rate approaches the optimum, it may take an intolerably long time before it is reasonably close. Thus, the stronger criterion sets a lower bound on the rate of approach to the optimum. For example, the criterion would designate a sequence $\langle c_T \rangle$ approaching 0 (such as $\langle (k/(T+k))^j \rangle$ or $\langle k/(k+e^{jT}) \rangle$, for $0 < j < \infty$) and then require for all T

$$[U_{\tau,E}(T)/U_E^*(T)] > (1 - c_T).$$

Clearly the plan τ satisfies the asymptotic optimal rate criterion when it satisfies this criterion and, in addition, τ can approach that rate no more slowly than c_T approaches 0.

The simplest way to extend these criteria to $\mathcal{E}$ is to require that a plan $\tau \in \mathfrak{J}$ meet the given criterion in each $E \in \mathcal{E}$.

τ is *robust in* $\mathcal{E}$ *with respect to the asymptotic optimal rate criterion for* $\mathfrak{J}$ when

$$\operatorname*{glb}_{E\in\mathcal{E}} \lim_{T\to\infty} [U_{\tau,E}(T)/U_E^*(T)] = 1.$$

τ is *robust in* $\mathcal{E}$ *with respect to the interim behavior criterion* $<c_T>$ *for* $\mathfrak{J}$ when, for all T,

$$\operatorname*{glb}_{E\in\mathcal{E}} [U_{\tau,E}(T)/U_E^*(T)] > (1 - c_T).$$

Each criterion in effect classifies the plans in $\mathfrak{J}$ as "good" or "bad" according to whether or not it is satisfied. The first of these criteria is commonly met in a wide range of applications, while the second proves to be relevant to questions of survival under competition. (Once again, a plan satisfying the second criterion automatically meets the first but not vice versa.) Other criteria can be based on the cumulative payoff function and indeed criteria of a quite different kind can be useful in particular situations. Nevertheless the criteria given are representative and of general use; they will play a prominent role later.

2. PRESENTATION

A problem in adaptation will be said to be well posed once $\mathfrak{J}$, $\mathcal{E}$, and χ have been specified within the foregoing framework. An *adaptive system* is specified within this framework by the set of objects $(\mathcal{A}, \Omega, I, \tau)$ where

$\mathcal{A} = \{A_1, A_2, \ldots\}$ is the set of attainable structures, the domain of action of the adaptive plan,

$\Omega = \{\omega_1, \omega_2, \ldots\}$ is the set of operators for modifying structures with $\omega \in \Omega$ being a function $\omega : \mathcal{A} \rightarrow \mathcal{P}$, where $\mathcal{P}$ is some set of probability distributions over $\mathcal{A}$,

I is the set of possible inputs to the system from the environment, and

$\tau : I \times \mathcal{A} \rightarrow \Omega$ is the adaptive plan which, on the basis of the input and structure at time t, determines what operator is to be applied at time t.

Under the intended interpretation

$$\tau(I(t), \mathcal{A}(t)) = \omega_t \in \Omega \quad \text{and} \quad \omega_t(\mathcal{A}(t)) = \mathcal{P}(t+1),$$

where $\mathcal{P}(t+1)$ is a particular distribution over $\mathcal{A}$. $\mathcal{A}(t+1)$ is determined by drawing a random sample from $\mathcal{A}$ according to the distribution $\mathcal{P}(t+1)$. Given the input sequence $\langle I(1), I(2), \ldots \rangle$, τ completely determines the stochastic process. (Occasionally, when the adaptive system is to be deterministic with $\mathcal{A}(t+1)$ being uniquely determined once $I(t)$ and $\mathcal{A}(t)$ are given, τ will be defined without the use of operators so that $\tau: I \times \mathcal{A} \rightarrow \mathcal{A}$.) The structure of the adaptive system at time t, $\mathcal{A}(t)$, will be required to summarize whatever aspects of the input history are to be available to the plan. Hence it will often be useful to represent $\mathcal{A}$ as $\mathcal{A}_1 \times \mathcal{M}$, where $\mathcal{A}_1$ is the set of structures to be directly tested and $\mathcal{M}$ is the set of possible memory configurations, for retaining past history not directly incorporated in the tested structures.

$\mathcal{J}$ is the set of feasible or possible plans of the form $\tau: I \times \mathcal{A} \rightarrow \Omega$ (or $\tau: I \times \mathcal{A} \rightarrow \mathcal{A}$) appropriate to the problem being investigated.

$\mathcal{E}$ represents the range of possible environments or, equivalently, the initial uncertainty of the adaptive system about its environment. When the plan τ tries a structure $\mathcal{A}(t) \in \mathcal{A}$ at time t, the particular environment $E \in \mathcal{E}$ confronting the adaptive system signals a response $I(t) \in I$. The performance or payoff $\mu_E(\mathcal{A}(t))$, given by the function $\mu_E: \mathcal{A} \rightarrow$ *Reals*, is generally an important part of the information $I(t)$. Given $E \neq E'$ for $E, E' \in \mathcal{E}$, the corresponding functions $\mu_E, \mu_{E'}$ are generally not identical so that a major part of the uncertainty about the environment is just about how well a structure will perform therein. When a plan employs, or receives, only information about payoff so that $I(t) = \mu_E(\mathcal{A}(t))$ it will be called a *payoff-only* plan.

Finally, the various plans in $\mathcal{J}$ are to be compared over $\mathcal{E}$ according to a criterion χ. Comparisons will often be based on the cumulative payoff functions $U_{\tau,E}(T) = \sum_{t=1}^{T} \bar{\mu}_E(\tau, t)$, where $\bar{\mu}_E(\tau, t)$ is the expected payoff under $\mathcal{P}(t)$, and the "target" function $U_E^*(T) = \operatorname{lub}_{\tau \in \mathcal{J}} U_{\tau,E}(T)$. An *interim behavior* criterion, based on a selected sequence $\langle c_T \rangle \rightarrow 0$ and of the form

$$\operatorname{glb}_{E \in \mathcal{E}} [U_{\tau,E}(T)/U_E^*(T)] > (1 - c_T),$$

will be important in the sequel.

With the help of this framework each of the fundamental questions about adaptation posed in chapter 1, section 1, can be translated into a formal counterpart:

Original	*Formal*
To what parts of its environment is the organism (system, organization) adapting?	What is $\mathcal{E}$?
How does the environment act upon the adapting organism (system, organization)?	What is I?
What structures are undergoing adaptation?	What is $\mathcal{A}$?
What are the mechanisms of adaptation?	What is Ω?
What part of the history of its interaction with the environment does the organism (system, organization) retain in addition to that summarized in the structure tested?	What is $\mathcal{M}$?
What limits are there to the adaptive process?	What is $\mathcal{J}$?
How are different (hypotheses about) adaptive processes to be compared?	What is χ?

3. COMPARISON WITH THE DUBINS-SAVAGE FORMALIZATION OF THE GAMBLER'S PROBLEM

> . . . much of the mathematical essence of a theory of gambling consists of the discovery and demonstration of sharp inequalities for stochastic processes . . . this theory is closely akin to dynamic programming and Bayesian statistics. In the reviewer's opinion, [*How to Gamble If You Must*] is one of the most original books published since World War II.
>
> M. Iosifescu. *Math. Rev. 38*, 5, Review 5276 (1969).

For those who have read, or can be induced to read, Dubins and Savage's influential book, this section (which requires special knowledge not essential for subsequent development) shows how to translate their formulation of the abstract gambler's problem to the present framework and vice versa. Briefly, their formulation is based on a progression of *fortunes* $f_0, f_1, f_2, \ldots$ which the gambler attains by a sequence of gambles. A *gamble* is naturally given as a probability distribution over the set of all possible fortunes F. The gambler's range of choice at any time t depends directly and only upon his current fortune f_t so that, as Dubins and Savage remark, the word "state" might be more appropriate than "fortune." The gambler's range of choice for each fortune f is dictated by the *gambling house* Γ. The *strategy* σ for confronting the house is a function which at each time t selects a gamble in Γ on the basis of the sequence or *partial history* of fortunes to that time $(f_0, f_1, \ldots, f_t)$. Finally the *utility* of a given fortune f to the gambler is specified by a utility function u. Thus an abstract gambler's problem is well posed when the objects (F, Γ, u) have been specified; the gambler's response to the problem is given by his strategy σ.

The objects of the Dubins-Savage framework can be put in a one-to-one correspondence with formally equivalent objects in the present framework. With the help of this correspondence any theorem proved in one framework can automatically be translated to a statement which is a valid theorem in the other framework. The relation between the intended interpretations of corresponding objects is in itself enlightening, but the real advantage accrues from the ability to transfer results from one framework to the other with a guarantee of validity.

The following table presents the formal correspondence with an indication of the intended interpretation of each formal object. In this table the superscript "*" on a set will indicate the set of all *finite* sequences (or strings) which can be formed from that set; thus F^* is the set of all partial histories.

Dubins-Savage	*Adaptive Systems*
F, *fortunes*	$\mathcal{A}_1$, basic structures (see τ below).
γ, a probability distribution over fortunes or a *gamble*.	P, a probability distribution over structures, i.e., $P \in \mathcal{P}$.
Γ, a function assigning a set of gambles to each $f \in F$, the *house*.	The (induced) function which assigns to each $A \in \mathcal{A}$ the set of distributions $\mathcal{P}_A = \{\omega(A), \omega \in \Omega\}$.
$\sigma: F^* \to \{\gamma\}$, a *strategy* which assigns to each *partial history* $p \in F^*$ a gamble $\Gamma(f)$, where f is the latest fortune in the sequence p.	$\tau: \mathcal{A} \to \mathcal{P}$, an *adaptive plan;* τ uses only the *retained history* $\mathfrak{M}$ in $\mathcal{A} = \mathcal{A}_1 \times \mathfrak{M}$, but τ has the same generality as σ if $\mathcal{A}_1 = F$ and $\mathfrak{M} = F^*$.
$u: F \to$ Reals, *utility*.	$\mu_E: \mathcal{A} \to$ Reals, performance.

As implied by their terminology, Dubins and Savage treat situations wherein the *expectation* for any strategy σ, given an initial fortune F, is less than F. That is, the strategies are operating in environments wherein continued operation makes degraded performance ever more likely. (This is similar to adaptation in an environment having only nonreplaceable resources, so that performance can only decline in the long run.) In contrast, the present work is primarily concerned with complex environments wherein performance can be permanently improved, if only the right information can be acquired and exploited. Despite the differences, or more likely because of them, theorems from one framework have interesting, and sometimes surprising, translations in the other framework.

3. Illustrations

The formal framework set out in chapter 2 is intended, first of all, as an instrument for uniform treatment of adaptation. If it is to be useful a wide variety of adaptive processes must fit comfortably within its confines. To give us a better idea of how the framework serves this end, the present chapter applies the framework in several different fields. It will repay the reader to skim through all of the illustrations on first reading, but he should skip without hesitation over difficult points on unfamiliar ground, reserving concentration for illustrations from familiar fields. Although each of the illustrations adds something to the substantiation of the framework no one of them is essential in itself to later developments. The interpretations, limited usually to one commonly used model per field, are of necessity largely informal, but two points can be checked in each case: (1) the facility of the framework in picking out and organizing the facts relevant to adaptation, and (2) the fit of established mathematical models within the framework.

1. GENETICS

> . . . genes act in many ways, affecting many physiological and morphological characteristics which are relevant to survival. All of these come together into the sufficient parameter "fitness" or selective value. . . . Similarly environmental fluctuation, patchiness, and productivity can be combined . . . in . . . [a] measure of environmental uncertainty. . . .
>
> Levins in *Changing Environments* (pp. 6–7)

The phenotype is the product of the harmonious interaction of all genes. The genotype is a "physiological team" in which a gene can make a maximum contribution to fitness by elaborating its chemical "gene product" in the needed quantity and at the time when it is needed in development. There is extensive interaction not only among the alleles of a locus, but also between loci. The main locale of

> these epistatic interactions is the developmental pathway. Natural selection will tend to bring together those genes that constitute a balanced system. The process by which genes are accumulated in the gene pool that collaborate harmoniously is called "integration" or "coadaptation." The result of this selection has been referred to as "internal balance." Each gene will favor the selection of that genetic background on which it can make its maximum contribution to fitness. The fitness of a gene thus depends on and is controlled by the totality of its genetic background.
>
> Mayr in *Animal Species and Evolution* (p. 295)

We have already looked at genetic processes at some length in the preliminary survey, so this illustration will be brief, mostly recapitulating the main points of the earlier discussion, but within the formal framework. Typically, only a certain range of basic structures, i.e., chromosomes, is admitted to studies in genetics, so that only a species, family, or other taxonomic grouping is involved. Still, in principle, one can study all possible variations, including variations in chromosome number and type. The range of the study will be primarily determined by the set Ω of genetic operators admitted, since the possible variants (genotypic and phenotypic) will be those produced by sequences of genetic operators from Ω. Familiar examples of genetic operators are mutation, crossover, inversion, dominance modification, translocation, and deletion (see the formal definitions given in chapter 6).

The genetic adaptive plan develops in terms of an everchanging population of chromosomes which, interacting with the environment, provides a concurrent sequence of phenotype populations. For many purposes, it is convenient to represent a population as a probability distribution over the set of genotypes $\mathcal{A}_1$, where the probability assigned to genotype $A \in \mathcal{A}_1$ is the fraction of the total population consisting of that genotype (cf. Crow and Kimura 1970). Thus the population at time t can be specified by $\mathcal{A}(t) \in \mathcal{A}$, where $\mathcal{A}$ is the set of distributions over $\mathcal{A}_1$. In very general terms, each element of the population is tested against the environment and is ranked according to its fitness—its ability to survive and reproduce. It's often useful to think of the environment E in terms of environmental niches, each of which can be exploited by an appropriate set of phenotypic characteristics. Then fitness μ_E becomes a function of the coadapted sets of alleles which produce these characteristics (see chapter 4). From this point of view the population $\mathcal{A}(t)$ can be looked upon as a reservoir of coadapted sets, preserving the history of past advances, particularly the environmental niches encountered.

Most mathematical models of genetic adaptation are based on very simple reproductive plans, where each individual allele σ_i is assigned a fitness $\mu_E(\sigma_i)$ and the fitness of any set of alleles $\{\sigma_1, \ldots, \sigma_m\}$ is taken to be the sum of the fitnesses of the alleles in the set,

$$\sum_{i=1}^{m} \mu_E(\sigma_i).$$

However, in general, the fitness of an allele depends critically upon the influence of other alleles (epistasis). The replacement of any single allele in a coadapted set may completely destroy the complex of phenotypic characteristics necessary for adaptation to a particular environmental niche. The genetic operators provide for the preservation of coadapted sets by inducing a "linkage" between adjacent alleles—the closer together a set of alleles is on a chromosome, the more immune it is to separation by the genetic operators. Thus a more realistic set of adaptive plans provides for emphasis of coadapted sets through reproduction, combined with application of the genetic operators to provide new candidates and test established coadapted sets in new combinations and contexts.

More formally, an interesting set of plans can be defined in terms of a two-phase procedure: First the number of offspring of each individual A in a finite population $\mathcal{A}(t)$ is determined probabilistically, so that the expected number of offspring of A is proportional to A's observed fitness $\mu_E(A)$. The result is a population $\mathcal{A}'(t)$ with certain chromosomes emphasized, along with the coadapted sets they contain. Then, in the second phase, the genetic operators from Ω are applied (in some predetermined order) to yield the new population $\mathcal{A}(t+1)$. One class of plans of considerable practical relevance can be defined by assuming that operator ω_i from Ω is applied to an individual $A \in \mathcal{A}'(t)$ with probability p_i (constant over time). It is easy to see that the efficiency of such a plan will depend upon the values of the p_i; it is perhaps less clear that once each of the p_i has a value within a certain critical range, the plan remains efficient, relative to other possible plans, over a very broad range of fitness functions $\{\mu_E, E \in \mathcal{E}\}$. In particular, if chromosomes containing a given linked set of alleles repeatedly exhibit above-average fitness, the set will spread throughout the population. On the other hand, if a linked set occurs by happenstance in a chromosome of above-average fitness, later tests will eliminate it (see chapters 6 and 7). It is this mode of operation (and others similar) which gives such plans robustness—the ability to discover complex combinations of coadapted sets appropriate to a wide variety of environmental niches.

Because of the central role of fitness, it is natural to discuss the efficiency and robustness of a plan τ in terms of the average fitnesses of the populations it produces. Formally, the average fitness in E of a finite population of genotypes $\mathcal{A}_\tau(t)$ produced by τ at time t is given by

$$\bar{\mu}_E(\tau, t) = \frac{\sum_{A \in \mathcal{A}_\tau(t)} \mu_E(A)}{M(\tau, t)},$$

where $M(\tau, t)$ is the number of individuals in $\mathcal{A}_\tau(t)$. If we take the ratio $\bar{\mu}_E(\tau, t)/\bar{\mu}_E(\tau', t)$, we have an indication of how close τ comes to "extinction" relative to τ' (in the sense that extinction occurs when the population produced by τ becomes negligible relative to the population produced by τ'). If we take the greatest lower bound of this ratio relative to some set of possible plans $\mathfrak{J}$, we have an indication of the worst that can happen to τ in E, relative to $\mathfrak{J}$ at time t. Continuing in this vein we get the following criterion for ranking plans as to robustness over $\mathcal{E}$:

$$\underset{E \in \mathcal{E}}{\text{glb}}\ \underset{t}{\text{glb}}\ \underset{\tau' \in \mathfrak{J}}{\text{glb}}\ \bar{\mu}_E(\tau, t)/\bar{\mu}_E(\tau', t).$$

In effect this criterion ranks plans according to how close they come to extinction under the most unfavorable conditions.

The fantastic variety of possible genotypes, the effects of epistasis, changing environments, and the difficulty of retaining adaptations while maintaining variability (genetic variance), all constitute difficulties which genetic processes must surmount. In terms of the $(\mathfrak{J}, \mathcal{E}, \chi)$ framework these are, respectively, problems of the large size of $\mathcal{A}$, the nonlinearity and high dimensionality of μ_E, the nonstationarity of μ_E, and the mutual interference of search and exploitation. The $(\mathfrak{J}, \mathcal{E}, \chi)$ framework enables the definition of concepts (chapters 4 and 5) which in turn (chapters 6 through 9) help explain how genetic processes meet these difficulties in times consistent with paleological and current biological observations.

Summarizing:

$\mathcal{A}$, populations of chromosomes represented, for example, by the set of distributions over the set of genotypes $\mathcal{A}_1$.

Ω, genetic operators such as mutation, crossover, inversion, dominance modification, translocation, deletion, etc.

$\mathfrak{J}$, reproductive plans combining duplication according to fitness with the application of genetic operators; for example, if each operator $\omega_i \in \Omega$ is applied to individuals with a fixed probability p_i, then the set of possible plans can be represented by the set

$$\{(p_1, \ldots, p_i, \ldots, p_b) \text{ where } 0 \leqq p_i \leqq 1\}.$$

$\mathcal{E}$, the set of possible fitness functions $\{\mu_E : \mathcal{A} \to \mathcal{R}\}$, each perhaps stated as a function of combinations of coadapted sets.

χ, comparison of plans according to average fitnesses of the populations produced; for example,

$$\operatorname*{glb}_{E\in\mathcal{E}}\ \operatorname*{glb}_{t}\ \operatorname*{glb}_{\tau'\in\mathfrak{I}}\ \bar{\mu}_E(\tau, t)/\bar{\mu}_E(\tau', t).$$

2. ECONOMICS

> The specification of how goods can be transformed into each other is called the *technology* of the model and the specification of how goods are transformed to satisfaction is called the *utility function*. Given this structure and some initial bundle of goods, the problem of optimal development is to decide at each point of time how much to invest and how much to consume in order to maximize utility summed over time in some suitable way.
>
> Gale in "A Mathematical Theory of Optimal Economic Development" *Bull. AMS 74*, 2 (p. 207)

One of the most important formulations of mathematical economics is the von Neumann technology. This technology can be presented (following David Gale 1968) in terms of a finite set of *goods* and a finite set of *activities*, where each activity transforms some goods into others. If the goods are indexed, then the goods available to the economy at any given time can be presented as a vector where the ith component gives the amount of the ith good. In the same way, the input to the jth activity and the resultant output can be given by a pair of vectors W_j and W_j', where the ith component of W_j specifies the amount of the ith good required by the activity, while the ith component of W_j' specifies the amount produced. An activity can be operated at various levels of effort so that, for instance, if the amount of input of each required good is doubled then the amount of output will be doubled. More generally, if the level of effort for activity j is c_j then the pair $(W_jc_j, W_j'c_j)$ specifies the input and output of the activity. If a mixture of activities is allowed, the overall technology can be specified as the set of pairs

$$\{(Wc, W'c) \ni c \in Q\}$$

where W and W' are matrices having the vectors W_j and W_j' as their respective jth columns, each c is a vector having the level of the jth activity as its jth component, and Q designates the set of admissible activity mixes (corresponding to the real constraints limiting the total activity at any time). A *program* for utilizing the technology is given by a sequence of activities $\langle c_t \rangle$ satisfying the intuitive "local" requirement that the total amount of each good required as input for the activities at time

TYPICAL ACTIVITIES:

	"*Coal Storage*"		"*Coal Mining*"		"*Tool Fabrication*"		. . .
	Input	*Output*	*Input*	*Output*	*Input*	*Output*	
	.	.	.	.	.	.	
Wood	0	0	0	0	1	0	
Coal	1	1	0	4	0	0	
Iron Ore	0	0	0	0	0	0	. . .
Steel	0	0	0	0	1	0	
Tools	0	0	1	0	0	1	

Matrix W (*Combining Activity Input Requirements*) and *Matrix W′* (*Combining Activity Outputs*); rows: Wood, Coal, Iron Ore, Steel, Tools

$$W = \begin{bmatrix} 1 & 0 & 0 & 1 & 0 & 1 & 0 & 0 \\ 0 & 1 & 0 & 1 & 2 & 0 & 0 & 0 \\ 0 & 0 & 1 & 1 & 1 & 0 & 0 & 0 \\ 0 & 0 & 0 & 0 & 1 & 1 & 0 & 0 \\ 0 & 0 & 0 & 0 & 0 & 0 & 1 & 1 \end{bmatrix} \qquad W' = \begin{bmatrix} 3/2 & 0 & 0 & 0 & 0 & 0 & 0 & 0 \\ 0 & 1 & 0 & 0 & 0 & 0 & 4 & 0 \\ 0 & 0 & 1 & 0 & 0 & 0 & 0 & 4 \\ 0 & 0 & 0 & 1 & 3 & 0 & 0 & 0 \\ 0 & 0 & 0 & 0 & 0 & 1 & 0 & 0 \end{bmatrix}$$

TYPICAL PRODUCTION SEQUENCE:

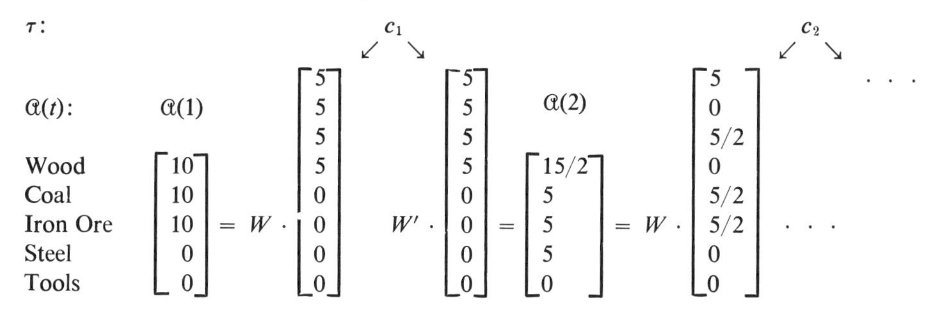

Fig. 3. Example of von Neumann technology

t cannot exceed the total amount of that good produced as output in the preceding period:

$$W'c_{t-1} \geqq Wc_t$$

(using matrix multiplication and the obvious extension of inequality to vectors). Activities which dispose of or store goods can be introduced so that the given inequalities can be changed to equalities without loss of generality. Thus, given an initial supply of goods $V(0)$, the set of admissible programs becomes

$$\mathcal{C} = \{\text{sequences } C_\beta = \langle c_{\beta,t} \rangle, \text{ where } \beta \in \mathcal{B}, \text{ an indexing set, and } t = 0, 1, 2, \ldots \ni \text{(i) } c_{\beta,t} \in Q, \text{ (ii) } Wc_{\beta,0} = V(0), \text{ (iii) } W'c_{\beta,t} = Wc_{\beta,t+1}\}.$$

It is assumed that each activity vector c can be assigned a unique *utility* $\mu(c)$ designating the satisfaction to society of engaging in the mix of activities specified by the vector. (This way of assigning utility has the nice feature that satisfying activities which do not directly consume goods, such as viewing pictures in a museum or conserving goods for future use, can be included in the model.) The object of the study is to compare various programs in terms of the utility sequences they produce. Typically, programs are compared over some interval of time $(0, T)$ by taking the difference of their accumulated utilities

$$U_\beta(T) - U_{\beta'}(T) \quad \text{where} \quad U_\beta(T) = \textstyle\sum_{t=0}^{T} \mu(c_{\beta,t}).$$

A program $C_{\beta*}$ is considered optimal if

$$\operatorname*{glb}_{C_\beta \in \mathcal{C}} \liminf_{T \to \infty} [U_{\beta*}(T) - U_\beta(T)] \geqq 0.$$

Because Q sets an upper limit to levels of effort, an optimal program always exists. A program C_β will often be satisfactory if its rate of accrual of utility $U_\beta(T)/T$ is comparable to that of $C_{\beta*}$.

Generally interest centers on "noncontracting" economies where, once an activity is possible, it continues to be possible at any subsequent time. This can be guaranteed, for example, if there is a set of initial goods which are "regenerated" by all activities (cf. sunlight, water, and air) and from which all other goods can be produced by appropriate sequences of activities. In such economies a mix of activities can be tried and, if found to be of above-average utility, can be employed again in the future.

In the $(\mathfrak{I}, \mathcal{E}, \chi)$ framework, the set of admissible activity mixes Q becomes the set of structures $\mathcal{A}$. An adaptive plan τ generates a program C by selecting a sequence of activity vectors $\langle c_{\tau,t} \rangle$ on the basis of information received from the environment (economy). The environment E in this case makes itself felt only through the observed utility sequence $\langle \mu_E(c_{\tau,t}) \rangle$; thus different utility functions correspond to different environments. Within this framework, the basic concern is discovery of an adaptive plan which, over a broad variety of environments, generates programs which work "near-optimally." A typical criterion of "near-optimality" would be that for all utility functions of interest the ratio of the rate of accrual of the adaptive plan τ, $U_\tau(T)/T$, to that of $C_{\beta^*}(E)$, $U_{\beta^*(E)}(T)/T$, approaches 1 for each $E \in \mathcal{E}$. That is,

$$\lim_{T\to\infty} [U_\tau(T)/U_{\beta^*(E)}(T)] = 1, \quad \text{for all } E \in \mathcal{E}.$$

Generally there will be some additional requirement that the rates be comparable for all times T.

Adaptation becomes important when there is uncertainty about just what utility should be assigned to given activity mixes, or when it is difficult to project μ_E into the future, or when Q is a function of time (reflecting technological innovations). The key to formulating an adaptive plan here, paralleling the procedure in other contexts, is continual use of incoming information (about satisfactions and dissatisfactions, changing technology, etc.) to modify activity levels. A well-formulated plan should respond automatically, specifying adjustments needed, as information accumulates. Since, in von Neumann's formulation, the environment is characterized by the utility assigned to different activity vectors, we can limit consideration to payoff-only plans. The fact that reproductive plans are payoff-only plans which can be proved near-optimal (in the sense defined above) for *any* set of utilities, makes it likely that such plans can supply the responsiveness required here. In $(\mathfrak{I}, \mathcal{E}, \chi)$ terms the basic problems here, as in the genetics illustration, are the large size of $\mathcal{A}$ coupled with nonlinearity and high-dimensionality of μ_E. Because the concepts of chapters 4 and 5 are formulated in terms of the general framework, they apply here as readily as to genetics. The resulting techniques are specifically interpreted as optimization procedures throughout chapter 6, at the end of section 7.1, and throughout section 7.2.

Summarizing:

$\mathcal{A}$, the set of admissible activity vectors Q.

Ω, transformations of Q into itself.

$\mathfrak{I}$, plans for selecting a program $\langle c_t \rangle \in \mathcal{C}$, where c_t is an activity vector in Q, on the basis of observed utilities $\{\mu_E(c_{t'}), t' < t\}$, i.e., payoff-only plans.

$\mathcal{E}$, an indexing set of possible utility functions $\{\mu_E: Q \to \mathcal{R}, E \in \mathcal{E}\}$.

χ, typically a requirement that, for all utility functions μ_E, $E \in \mathcal{E}$, the limiting rate of accrual of a plan, $\lim_{T\to\infty} (U_\tau(T)/T)$, equal that of the best possible program $C_{\beta^*}(E)$ in each $E \in \mathcal{E}$.

3. GAME-PLAYING

> Lacking such knowledge [of machine-learning techniques], it is necessary to specify methods of problem solution in minute and exact detail, a time-consuming and costly procedure. Programming computers to learn from experience should eventually eliminate the need for much of this detailed programming effort.
>
> Samuel in "Some Studies in Machine Learning Using the Game of Checkers" *IBM J. Res. Dev. 3* (p. 211)

Most competitive games played by man (board games, card games, etc.) can be presented in terms of a *tree* of moves where each *vertex* (point, node) of the tree corresponds to a possible game configuration and each directed *edge* (arrow) leading from a given vertex represents a legal move of the game. The edge points to a new vertex corresponding to a configuration which can be attained from the given one in one move (turn, action); the options open to a player from a given configuration are thus indicated by the edges leading from the corresponding vertex. The tree has a single distinguished vertex with no edges leading into it, the *initial* vertex, and there are *terminal* vertices, having no edges leading from them, which designate outcomes of the game. In a typical two-person game which does not involve chance, the first player selects one of the options leading from the initial configuration; then the second player selects one of the options leading from the resulting configuration; the *play* of the game proceeds with the two players alternately selecting options. The result is a *path* from the initial vertex to some terminal vertex. The outcomes are ranked, usually by a *payoff* function which assigns a value to each terminal vertex.

In these terms, a *pure strategy* for a given player is an algorithm (program, procedure) which, for each nonterminal configuration, selects a particular option leading therefrom. Once each player chooses a pure strategy, the outcome of the game is completely determined, although in practice it is usually possible to determine this outcome only by actually playing the game. Thus, in a strictly determined (non-chance) two-person game, each pair of pure strategies (one for each player) can be assigned a unique payoff. The object of either player, then, is to find a strategy which does as well as possible against the opponent as measured by the expected payoff. This informal object ramifies into a whole series of cases, depending upon the initial information about the opponent and the form of the game.

One of the simplest cases occurs when it is known that the opponent, say the second player, has selected a single pure strategy for all future plays of the game.

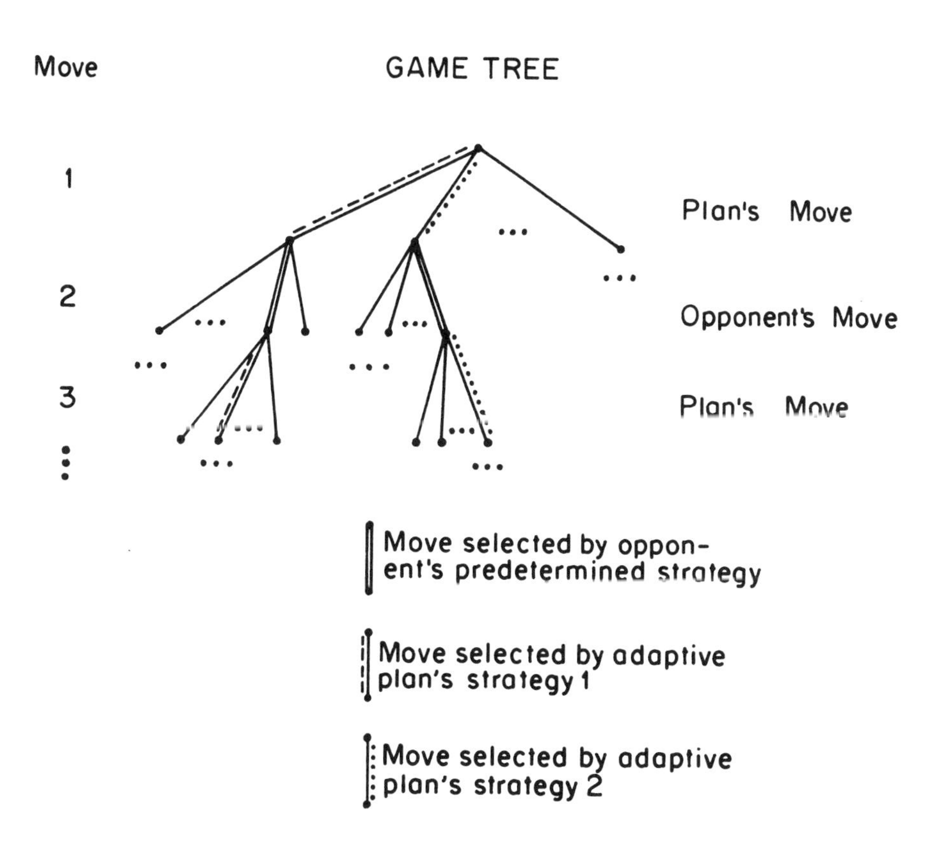

Fig. 4. Example of a game tree

The object of the first player, then, is to learn enough of the strategy chosen by the second player to find an opposing strategy which maximizes payoff. When the game tree involves only a finite number of vertices, as is often the case, it is at least theoretically possible to locate the maximizing strategy by enumerating and testing all strategies against the opponent. However, if there is an average of k options proceeding from each configuration, and if the average play involves m moves, there will be in excess of k^m pure strategies. The situation is quite comparable to the examples of enumeration given earlier. Even for a quite modest game with $k = 10$ and $m = 20$, and a machine which tests strategies at the exceptional rate of one every 10^{-9} second, it would require in excess of 10^{11} seconds, or about 30 centuries, to test all possibilities. Efficiency thus becomes the critical issue, and interest centers on the discovery of plans which enable a player to do well while learning to do better. If plans are compared in terms of accumulated payoff, a criterion emerges analogous to the classical "gambler's ruin" of elementary probability. Let $U_E(\tau, t)$ be the payoff accumulated to time t by plan $\tau \in \mathfrak{J}$ confronting the (unknown) pure strategy $E \in \mathcal{E}$, and require that

$$\operatorname*{glb}_{E \in \mathcal{E}} \operatorname*{glb}_{\tau \in \mathfrak{J}} \operatorname*{glb}_{t} \left[U_{\tau_1,E}(t) / U_{\tau,E}(t) \right] \geqq c.$$

That is, the payoff accumulated by τ_1 never falls to less than c of that accumulated by any other admissible plan $\tau \in \mathfrak{J}$, no matter what strategy the opponent chooses (even if that other plan by happenstance hits upon a good opposing strategy in its first trial). The smaller c is, the less stringent the criterion and, in general, the larger the number of plans satisfying the criterion. The usefulness of this criterion and the kinds of plans satisfying it will be discussed at length later (see especially the discussion of Samuel's algorithm in section 7.3); for now it is sufficient to notice that: (i) the criterion depends upon the accumulation function $U_{\tau,E}(t)$, (ii) for a given opposing strategy E, the lower the efficiency of a plan in accumulating payoff in relation to other plans $\tau \in \mathfrak{J}$, the smaller c becomes, and (iii) the rating of a plan will be determined by its performance against the opposing strategy which gives it the most difficulty.

Even when it is known that the opponent has selected a single pure strategy, there is a wide range of sophistication of adaptive plans. One class of simpler plans, the payoff-only plans, proves to be quite instructive because it sets a nontrivial lower bound on the performance of more sophisticated plans and it can be analyzed in some detail. In this context, a payoff-only plan ranks strategies it has tried according to the payoff obtained, and it generates new trial strategies on the basis of (selected parts of) this information alone (see section 7.3). More sophisticated plans use the large amounts of information generated during plays of the game, information concerning configurations encountered and the sequence in which they occur (see chapter 8). Obviously a plan which makes proper use of this additional information should do no worse than a payoff-only plan (since the sophisticated plan can reduce its operation to that of a payoff-only plan by ignoring the additional information), and there are certainly situations in which the information will enable the plan to accumulate payoff at a greater rate than a payoff-only plan.

The other extreme from a fixed opposing pure strategy occurs when any sequential mix of strategies is presumed possible on the part of the opponent. The object then (following von Neumann 1947) is usually to minimize the maximum loss (negative of the payoff) the opponent can impose. It is interesting that often (checkers, chess, go) this minimax strategy is a pure strategy. Thus, although the payoff may vary on successive trials of the same strategy, the plan can still restrict its search to pure strategies in such cases. In more general situations, however, the plan will have to employ stochastic mixtures of pure strategies and, if it is to exploit its opponents maximally, it will even associate particular mixtures with particular kinds of opponents (assuming it is supplied with enough information to enable it to identify individual opponents).

Considered in the $(\mathfrak{T}, \mathcal{E}, \chi)$ framework, the strategies become the elements of the domain of action $\mathcal{A}$ and the plans for employing these strategies become elements of $\mathfrak{T}$. The set of admissible environments $\mathcal{E}$ depends upon the particular case considered. If it is known that the opponent has chosen a single pure strategy, then the set of admissible environments $\mathcal{E}$ is given by the set of pure strategies. The criterion χ for ranking the plans is then built up from the unique payoff determined by each pair of opposing pure strategies, the example given being the "gambler's ruin" criterion

$$\operatorname*{glb}_{E \in \mathcal{E}} \operatorname*{glb}_{\tau \in \mathfrak{T}} \operatorname*{glb}_{t} [U_{\tau 1,E}(t)/U_{\tau,E}(t)].$$

In the more complicated cases, the set of environments is enlarged, ultimately including plans over $\mathcal{A}$; however, the accumulation functions $U_{\tau,E}(t)$ are still defined and criteria such as the "gambler's ruin" criterion can still be used to rank the plans in $\mathfrak{T}$.

Once again, as in the previous two illustrations, the large size of $\mathcal{A}$ and the complex relation of its elements to performance constitute a major barrier to improvement. Section 7.3 specifically discusses the role of adaptive algorithms in game strategy spaces defined in the manner of Samuel. In addition, the necessity of using non-payoff information generated *during* the play of more complex games presents special difficulties. This latter problem is addressed in section 8.4 as an elaboration of the concepts and techniques developed in the earlier chapters.

Summarizing:

$\mathcal{A}$, strategies for the game.

Ω, dependent upon the way strategies are represented; genetic operators will function if descriptors are used so that each strategy is designated by a string of descriptor values (see the predictive modeling technique of the next section for suggestions concerning operations on the strategy *during* the play; section 8.4 extends these ideas).

$\mathfrak{T}$, plans for testing strategies.

- $\mathcal{E}$, the strategic options open to the opponent; in simple cases, the set of pure strategies.
- χ, a ranking of plans using the cumulative payoff functions, the "gambler's ruin" criterion being an example.

4. SEARCHES, PATTERN RECOGNITION, AND STATISTICAL INFERENCE

Searches occur as the principal element in most problem-solving and goal-attainment attempts, from maze-running through resource allocation to very complicated planning situations in business, government, and research. Games and searches have much in common and, from one viewpoint, a game is just a search (perturbed by opponents) in which the object is to find a winning position. The complementary viewpoint is that a search is just a game in which the moves are the transformations (choices, inferences) permissible in carrying out the search. Thus, this discussion of searches complements the previous discussion of games.

In complicated searches the attainable situations $\mathcal{S}$ are not given explicitly; instead some initial situation $S_0 \in \mathcal{S}$ (position in a maze, collection of facts, etc.) is specified and the searcher is given a repertory of transformations $\{\eta_i\}$ which can be applied (repeatedly) in carrying out the search. As in the case of games, a tree is a convenient abstract representation of the search. For searches, each edge corresponds to a possible transformation η_i and the traverse of any path in the tree corresponds to the application of the associated sequence of transformations. The vertex at the end of a path extending from the initial vertex corresponds to the situation produced from the initial situation by the transformations associated with the path. The difficulty of solving a problem or attaining a goal is primarily a function of the size of the search tree and the cost of applying the transformations. In most cases of interest the trees are so vast that hope of tracing out all alternative paths must be abandoned. Somehow one must formulate a search plan which, over a wide range of searches, will act with sufficient efficiency to attain the goal or solve the problem.

A typical search plan (see Newell, Shaw, and Simon's [1959] GPS or Samuel's [1959] procedure) involves the following elements:

(i) An (ordered) set of feature detectors $\{\delta_i : \mathcal{S} \to V_i$, where V_i is the range of readings or outputs of the ith detector$\}$. Typically, each detector is an algorithm which, when presented with a "scene" or situation, calculates a number; if the number is restricted to 0 or 1, it is convenient to think of the algorithm as detecting the presence or absence of a property (cf. the simple artificial adaptive system of section 1.3). The need for detectors arises from the overwhelming flow of information in most realistic situations; the intent is to filter out as much "irrelevant" information as possible.

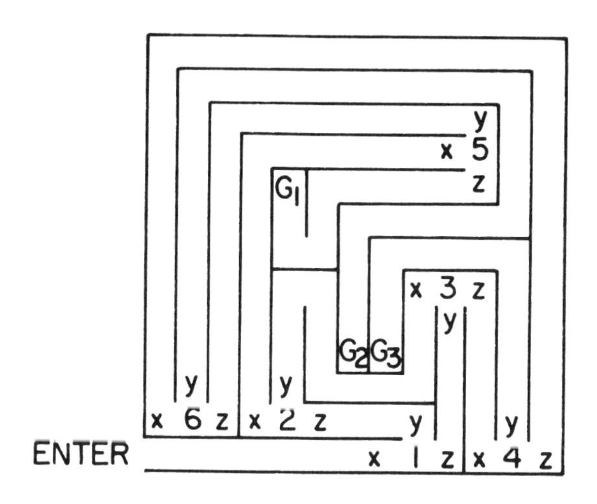

At each choice point, 1 through 6, there is to be a sign associated with each of the 3 possible directions x, y, z. If the symbol "$\wedge$" occurs at the top of a sign the associated corridor belongs to the shortest path from the entrance to the goal; on the other hand, if the symbol "$\vee$" occurs at the bottom of a sign the associated corridor is to be avoided. Either symbol may be dark on a light background or vice versa. Thus, reduced to a 4-by-4 array of sensors (see section 1.3), either of the configurations

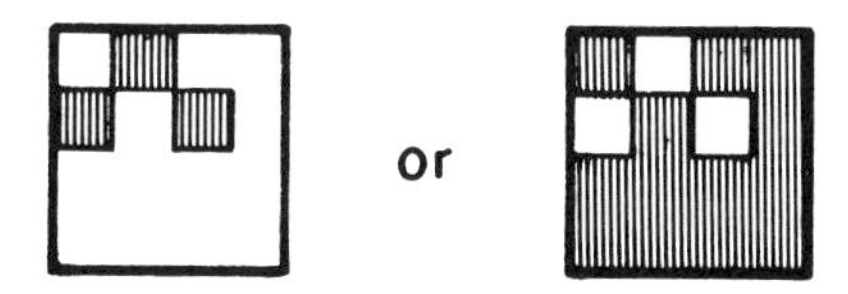

indicates the direction of the goal. Each experiment involves a set of signs indicating uniquely the shortest path to one of the three possible goals G_1, G_2, G_3.
In the terminology of section 3.4, the state at each choice point is given by the triple of signs there. That is,

$$S = \{\text{triples of 4-by-4 arrays}\}$$

$$\{\eta_1,\eta_2,\eta_3\} = \{\text{"follow direction } x\text{," "follow direction } y\text{," "follow direction } z\text{"}\}$$

Fig. 5. A simple search setting: a maze with six choice points

The device is supplied with four detectors (see section 1.3)

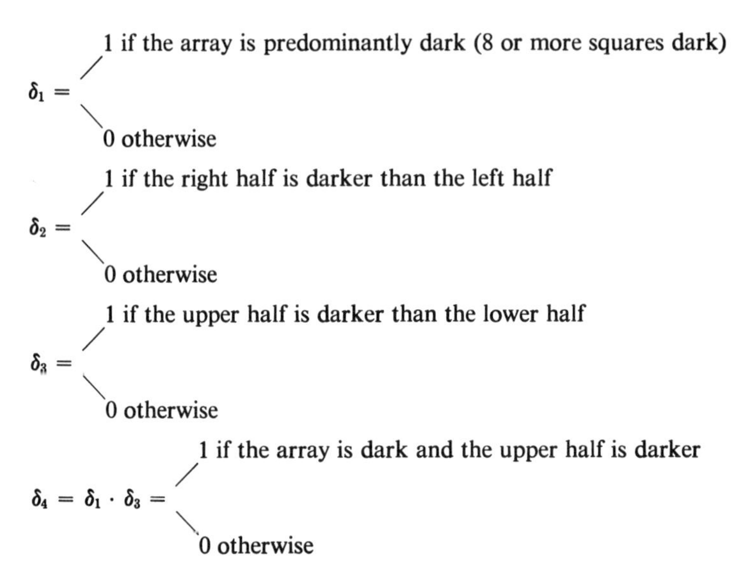

Thus the array

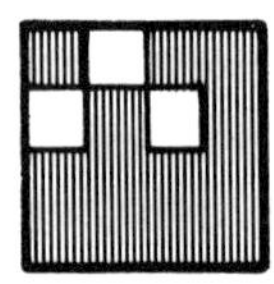

would be assigned the quadruple of detector values (1,1,0,0). (Note the large reduction in the number of situations to be evaluated—there are $2^{16} \cong 64{,}000$ different arrays but only $2^4 = 16$ detector value quadruples.)

The threshold devices of interest are specified by

$$f(S) = \sum_{i=1}^{4} w_i \delta_i(S)$$

where each w_i can be chosen from the set $\{-2,-1,0,1,2\}$. A 4-by-4 array S is assigned to C^+ if $f(S) \geqq \frac{1}{2}$, otherwise it is assigned to C^-.

To determine what transformation from $\{\eta_1,\eta_2,\eta_3\}$ is to be invoked at choice point j each of the three arrays S_{jx}, S_{jy}, S_{jz} is submitted to f. If exactly one of S_{jx}, S_{jy}, S_{jz} is assigned to C^+ the corresponding corridor is followed, otherwise a corridor is chosen at random (and, presumably, the adaptive plan is invoked to modify the weights because of the lack of a unique prediction).

Fig. 6. A threshold device for the setting of figure 5

Setting I: The goal is at G_1, the signs at choice point 1 are

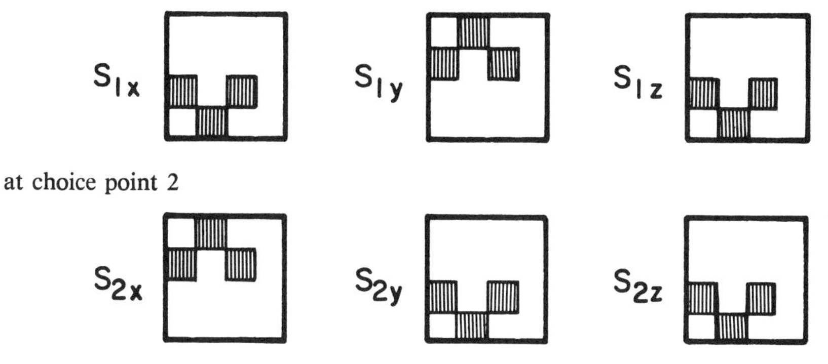

at choice point 2

and so on (i.e., the shortest path is indicated by dark symbols on a light background).

Setting II: The goal is at G_3, the signs at choice point 1 are

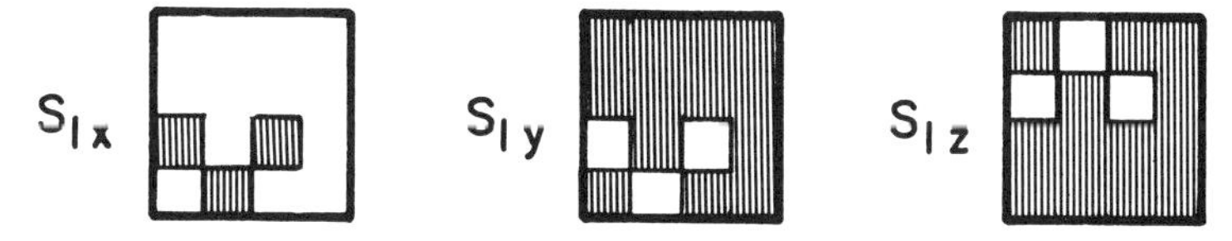

and at choice point 2 they are the same as in Setting I except for

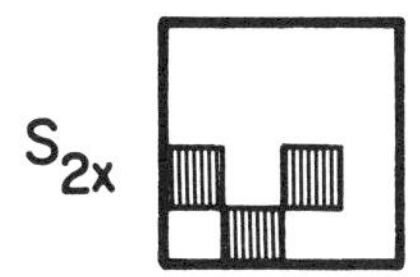

If the shortest path to the goal were always indicated as in Setting I, i.e., with dark symbols on a light background, then the function $f'(S) = \delta_3(S)$ (i.e., $w_3 = 1$, $w_1 = w_2 = w_4 = 0$) would always suffice for following the path. Notice, however, that in Setting II f' assigns exactly the same set of values (0,1,0) at point 1, indicating that f' does not distinguish the two settings. But, in Setting I f' assigns (1,0,0) at point 2, while in Setting II f' assigns (0,0,0) at point 2. Thus, starting from the same initial state (0,1,0) and invoking the same response η_y, f' arrives at two *different* states. Changing the weight assigned to δ_3 cannot correct the difficulty. This is a clear indication that the set of detectors (δ_3 in this case) is inadequate.

A quick check of the possibilities shows that consistently correct choices in the two settings can be achieved only by assigning a nonzero weight to δ_4, which is a nonlinear combination of δ_1 and δ_3. The function $f''(S) = \delta_1 + \delta_3 - 2\delta_4$ then performs correctly in both settings and, in fact, performs consistently with any proper sequence of signs.

Fig. 7. Some searches using the devices of figure 6 in the settings of figure 5

(ii) An evaluator. The evaluator calculates an estimate of the "distance" of any given situation from the goal, using the detector outputs (an ordered set of real numbers) produced by that situation. The estimates are supposed to take the costs of the transformations, etc., into account; that is, the "distances" are usually weighted path lengths, where the paths involved are (conjectured) sequences of transformations leading from given situations to the goal. The intent is to use these estimates to determine which transformations should be carried out next. An evaluation is made of each of the situations which could be produced from the current one by the application of allowed (simple sequences of) transformations, and then that (sequence of) transformation(s) is executed which leads to the new situation estimated to be "nearest" the goal.

(iii) Error correction procedures. Before the search plan has been tried, the detectors and evaluator must be set up in more or less arbitrary fashion, using whatever information is at hand. The purpose of the error correction procedures is to improve the detectors and evaluators as the plan accumulates data. The shorter term problem is that of evaluator improvement. A typical procedure is to explore the search tree to some distance ahead of the current situation, either actually or by simulation, evaluating the situations encountered for their estimated distances from the goal. The evaluation of the situation estimated to be "nearest" the goal is then compared with the evaluation of the current situation and the evaluator is modified to make the estimates consistent. This "lookahead" procedure decreases the likelihood of contradictory distance estimates at different points on the same path. (A similar procedure can be carried out without lookahead using predictors to make predictions about future situations, subsequently modifying the predictors to bring predictions more in line with observed outcomes.) As a result, the consistency of the evaluator is improved with each successive evaluation. At the same time, in most searches, the difficulty of estimating the distance to the goal decreases as the goal is approached, becoming perfect when the lookahead actually encounters the goal. Thus increasing the consistency ultimately increases the relevance of the evaluator.

There is, however, a caveat. If the set of detectors is inadequate, for whatever reason, the improvement of the evaluator will be blocked. This raises the broad issue of pattern recognition, for the set of detectors is, of course, meant to enable the plan to recognize critical features for goal-attainment. The plan must be able to classify each situation encountered according to the goal-directed transformation which should be applied to it. The long-term problem is that of determining whether the set of detectors is adequate to this task. Important shortcomings are indicated when, from application of identical transformations to situations classed as equivalent by the detectors, situations with critically different evaluations result. When this happens, the detectors have clearly failed to distinguish some feature which makes a critical difference as far as the transformations are concerned. The object, then, is to generate a detector which gives different readings for the previously indistinguishable situations. Among the obvious candidates are modifications of the detectors which made the distinctions *after* the transformations were applied. Usually simple modifications will enable such detectors to make the distinction *before* the transformation as well as *after*.

We can look at this whole problem in another way, a way which makes contact with standard definitions in the theory of probability. Assume that the search plan assigns to each transformation η a probability dependent upon the observed situation. That is, if S_α is the current situation, then each situation $S_\beta \in \mathcal{S}$ can be assigned a conditional probability of occurrence $p_{\alpha\beta}$, where $p_{\alpha\beta}$ is simply the sum of the probabilities of all transformations leading from S_α to S_β. (It may, of course, be that there are no transformations of S_α to S_β, in which case $p_{\alpha\beta} = 0$.) A sequence of trials performed according to the probabilities $p_{\alpha\beta}$ is a *Markov chain*, the outcome of each trial being a *random variable* (dependent upon the outcome of prior trials). The *sample space* underlying this random variable is the set of situations $\mathcal{S}$. Let us assign a measure of utility or relevance to each of these situations. (For example, goals could be assigned utility 1 and all other situations utility 0, or some more complicated assignment ranking goals and intermediate situations could be used.) Then, formally, the function W making this assignment is also a random variable. Accordingly, we can assign an expected utility to the random variable representing the outcome of each trial in the Markov chain. In these terms, the plan continually *redefines* the Markov chain (by changing the transformation probabilities). It attempts in this way to increase the average (over time) of the expected values of the sequence of random variables corresponding to its trials.

The role of detectors here is, as already suggested, reduction of the size of the sample space and simplification of the search. More formally, consider a set of n detectors (not necessarily all those available), $H = \{\delta_1, \ldots, \delta_n\}$, where H is arbitrarily ordered. The detectors in H assign to each $S \in \mathcal{S}$ an n-tuple of readings $(v_1, \ldots, v_n)$ belonging to the direct product

$$\Pi_{i=1}^{n} V_i.$$

In general there will be many situations producing a given set of detector readings; let $\mathcal{S}(v_1, \ldots, v_n)$ be the set of situations in $\mathcal{S}$ producing the particular n-tuple of readings $(v_1, \ldots, v_n)$. In probabilistic terms, $\mathcal{S}(v_1, \ldots, v_n)$ is an *event* defined on the sample space $\mathcal{S}$. Events themselves can be treated as random variables. (In fact, an occurrence of the situation S can be construed as the occurrence of all the events of which it is an instance.) Moreover, the function W assigning values to elements of $\mathcal{S}$ can be restricted to the event $\mathcal{S}(v_1, \ldots, v_n)$ so that it becomes a random variable $W(v_1, \ldots, v_n)$ over $\mathcal{S}(v_1, \ldots, v_n)$. As such $W(v_1, \ldots, v_n)$ has a well-defined expected value $\overline{W}(v_1, \ldots, v_n)$ over $\mathcal{S}(v_1, \ldots, v_n)$.

This probabilistic view of search plans is closely related to statistical inference based on sampling plans. The estimation of $\overline{W}(v_1, \ldots, v_n)$ from observation of a few samples drawn from $\mathcal{S}(v_1, \ldots, v_n)$ is a standard problem of statistical inference. We can think of a subset of detectors H as detecting one kind of critical feature when the corresponding $\overline{W}(v_1, \ldots, v_n)$ is greater than $\overline{W}$, where $\overline{W}$ is the average value of the random variable W over the sample space $\mathcal{S}$. Search plans go further in attempting to infer something of the value of $\overline{W}(v_1, \ldots, v_n)$ for $\mathcal{S}(v_1, \ldots, v_n)$ which have not been sampled. For example, $\mathcal{S}(v_1, v_2)$ is contained in both $\mathcal{S}(v_1)$ and $\mathcal{S}(v_2)$; often it is possible to infer something of $\overline{W}(v_1, v_2)$ from knowledge of $\overline{W}(v_1)$ and $\overline{W}(v_2)$, though not necessarily by standard statistical techniques.

The earlier concern with distinguishability is also directly stated in these terms: Let $\delta(t)$ be the particular n-tuple of detector readings at time t $(\delta(t) \in \Pi_i V_i)$ and let $f: \Pi_i V_i \rightarrow \{\eta\}$ be a search plan. That is, f is a prescription which specifies, for each set of detector readings, a transformation. The object of the search plan is to transform the current situation into one of high utility. But, for this to be possible, the effects of the transformations must be reliably indicated by the detectors. In particular, consider S_1 and $S_2 \in \mathcal{S}(v_1, \ldots, v_n)$, so that at $t = 1$ either would show the same reading $\delta(1) = (v_1, \ldots, v_n) \in \Pi_i V_i$. The plan f specifies the action $\eta(1) = f(\delta(1))$, and this in turn produces a new detector reading $\delta(2)$. The whole procedure is iterated to yield a *sequence* of pairs $\langle[\delta(1), f(\delta(1))], [\delta(2), f(\delta(2))], \ldots, [\delta(t), f(\delta(t))]\rangle$. The requirement on distinguishability is simply that, using the information provided by the detectors, f *reliably* transforms S_1 and S_2 into situations S_1' and S_2', respectively, for which $W(S_1') \cong W(S_2')$. (Notice that this is a much weaker requirement than would be necessary for a completely "autonomous" model wherein future situations would be wholly predictable on the basis of $\delta(1)$ without any further information from the environment. That is, in an autonomous model, knowledge of $\delta(1)$ and $\eta(1), \ldots, \eta(t)$ must suffice to determine $\delta(t + 1)$. This requirement for "autonomy"—technically a requirement that the detectors induce a homomorphism—can be quite difficult to meet and, for intricate

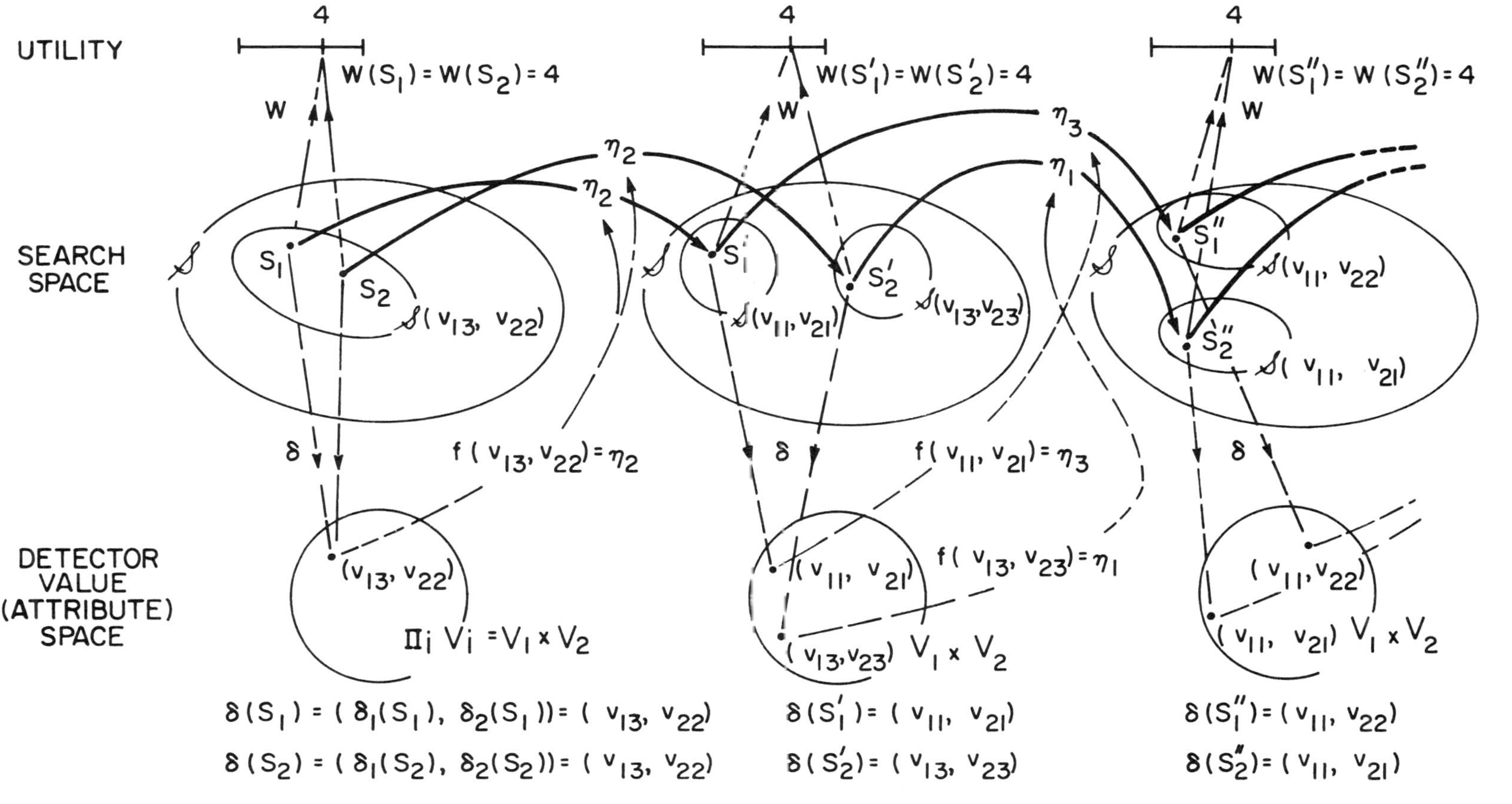

Fig. 8. A utility-consistent model using detector outputs to determine successive transformations

situations, there may be no nontrivial homomorphisms.) The requirement on distinguishability permits the detectors to lump together situations having approximately equal expected utilities. This permits us to construct new, much smaller Markov chains based on events of interest. With this interpretation, the object of the adaptive plan τ is to test different sets of detectors, H, and search plans, f.

The error correction procedures for the detectors and the evaluator can be coordinated via a model of the search environment. In complex search environments, it is the detectors which make a model possible. From considerations of storage alone, a model would be impossible if, for each observed transformation, the initial situation and its successor had to be recorded in full. If the system records just the effect of transformations on the detector readings, it can be reduced to manageable proportions. Stated another way, the state space of the search environment is reduced to the manageable space of detector readings, and the effects of the transformations are observed on this reduced space. The construction of the model proceeds as data accumulate. When new detectors are required to increase distinguishability, an augmented model can be built around the old model as a nucleus. In particular, when a new detector is added to the set of detectors, this does not affect the data or the part of the model concerned with the old set of detectors. The task is to add information about the effect of the transformations upon the new detector, particularly in those situations that were causing difficulty. Once a model is available, it can be used with the evaluator to generate predictions and these can be checked against the outcome of that segment of the search. The resulting error indications, together with simulated lookahead, can then be used to improve the consistency of the evaluator.

The sophistication of a model-evaluator search plan can only be justified when repeated searches must be made in the same overall search environment. As a prototype, we can consider an environment which (i) is complicated enough to make the exact recurrence of any particular situation extremely unlikely, but (ii) is regular enough to exhibit critical features (patterns) "pointing the way" to goals. The object of a plan, then, is to search out a succession of goals, improving its performance by incorporating the critical features in its detector-evaluator scheme.

The overall objective of a formal study of this area is to find a plan which, when presented with any of a broad range of complex search environments, rapidly increases its search efficiency by extracting and exploiting critical features. Considered within the $(\mathfrak{T}, \mathcal{E}, \chi)$ framework, the domain of action $\mathcal{A}$ of a search plan $\tau \in \mathfrak{T}$ consists of the various combinations of detectors, models, and evaluators that the plan can generate. Usually these will all be specified as algorithms or programs in some common formal language (see chapter 8). A search plan $\tau \in \mathfrak{I}$ thus amounts to a data-dependent algorithm for modifying the combination of detectors, model, and evaluator along the lines indicated above. The outward effect, at each point in time, of the combination produced by the search plan is a transformation η in the search environment. The range of the plan's action at each moment is therefore circumscribed by the set of possible transformations $\{\eta\}$. The set of admissible environments $E \in \mathcal{E}$ consists of the set of search environments over which the search plan is expected to operate, each element E being presentable as a tree generated by the possible transformations. Let $U_{\tau,E}(t)$ be the cost in E of the transformations applied by τ through time t. If $n_{\tau,E}(t)$ is the number of goals achieved to time t, then $c_{\tau,E}(t) = U_{\tau,E}(t)/n_{\tau,E}(t)$ is the average cost to time t of each goal in E. A conservative measure of a plan's performance over all of $\mathcal{E}$ would then be

$$\operatorname*{lub}_{E \in \mathcal{E}} \lim_{t \to \infty} c_{\tau,E}(t)$$

which yields the criterion χ wherein plan τ has a higher rank than plan τ' if it is assigned a lower number by the above measure. Suggestions for a model-evaluator plan, based on the genetic algorithms of chapter 6 and capable of modifying its representations, are advanced in section 8.4.

Summarizing:

$\mathcal{A}$, [probabilistic] Markov chains induced by the sets of conditional probabilities $\{p_i(S)$, the probability of applying transformation η_i to situation $S \in \mathcal{S}\}$; [general] admissible detector-evaluator-model combinations.

Ω, [probabilistic] rules for modifying the conditional probabilities $\{p_i(S)\}$; [general] "lookahead" error correction, detector generation, and model revision procedures.

$\mathfrak{I}$, algorithms for applying operators from Ω to $\mathcal{A}$ using information about the (sampled) average cost of goal attainment and (in the general case) errors in prediction ("lookahead") and observed inadequacies in detectors.

$\mathcal{E}$, the set of search environments characterized by search trees along with a transformation cost function $\mu_E(\eta_i, S)$ giving the cost of applying η_i in situation $S \in \mathcal{S}$.

χ, the ranking of plans in $\mathfrak{I}$ according to performance measures such as

$$\operatorname*{lub}_{E \in \mathcal{E}} \lim_{t \to \infty} c_{\tau,E}(t).$$

5. CONTROL AND FUNCTION OPTIMIZATION

> The fact that we need time to determine the minimum of the [performance] functional or the optimal [control] vector c^* is sad, but unavoidable—it is a cost that we have to pay in order to solve a complex problem in the presence of uncertainty. . . . adaptation and learning are characterized by a sequential gathering of information and the usage of current information to eliminate the uncertainty created by insufficient *a priori* information.
>
> Tsypkin in *Adaptation and Learning in Automatic Systems* (p. 69)

In the usual version, a controlled process is defined in terms of a set of variables $\{x_1, \ldots, x_k\}$ which are to be controlled. (For example, a simple process of air conditioning may involve three critical variables, temperature, humidity, and air flow.) The set of states or the *phase space* for the process, X, is the set of all possible combinations of values for these variables. (Thus, for an air conditioning process the phase space would be a 3-dimensional space of all triples of real numbers (x_1, x_2, x_3) where the temperature x_1 in degrees centigrade might have a range $0 \leqq x_1 \leqq 50$, etc.) Permissible changes or *transitions* in phase space are determined as a function of the state variable itself and a set of control parameters C. Typically X is a region in n-dimensional Euclidean space and the control parameters assume values in a region C of an m-dimensional space. Accordingly, the equation takes the form of a "law of motion" in the space X,

$$dx/dt = f(X(t), C(t)), \quad \text{where } X(t) \in, X\ C(t) \in C.$$

Often X will have several components $X_1, \ldots, X_k$ following distinct laws $f_1, \ldots, f_k$ so that

$$f(X(t), C(t)) = (f_1(X_1(t), C(t)), \ldots, f_k(X_k(t), C(t))).$$

For example, given a pursuit problem with a moving target having coordinates $X_2(t)$ at time t, $f_2(X_2(t), C(t))$ would be the law of motion of the target while $f_1(X_1(t), C(t))$ would determine the pursuit curve. If some component, say X_3, represents time, then $f_3(X_3(t), C(t)) = t$ and the law of motion becomes an explicit function of time.

When a rule or *policy* A is given for selecting elements of C as a function of time, a unique trajectory $\langle(X(t), C(t))\rangle$ through $X \times C$ is determined by the law of motion f. The object is to select a policy A for minimizing a given function J which assigns a performance or cost to each possible trajectory $\langle(X(t), C(t))\rangle$. In practice, the function J is usually determined as the cumulation over time of some instantaneous cost rate $Q(X(t), C(t))$; i.e., $J(\langle(X(t), U(t))\rangle) = \int Q(X(t), C(t))\,dt$. Typically, the cost function is derived from an explicit control objective such as attainment of a target state or a target region in minimal time or minimization of cumulative error. (Error is defined in terms of a measure of distance imposed on the phase space; the distance of the current state from the target region is the current error.) Control is thus a continuing search in phase space for the (usually moving) target or goal—as such the considerations of the preceding illustration are directly relevant. In the formulation of the pursuit problem stated above a natural measure of the cost of pursuit over some interval T would be the change in distance between target and pursuer divided by the fuel expenditure (with suitable conventions for trajectories where the distance does not decrease).

Although the controlled process is defined above in terms of continuous functions, discrete finite-state versions closely approximating the continuous version almost always exist. Indeed, if the problem is to be solved with the help of a digital computer, it must be put in finite-state form. Because the framework we are using is discrete, we will reformulate the problem in discrete form. The law of motion is given by

$$X(t+1) = f(X(t), C(t)),$$

and the cumulative cost for a given trajectory over T units of time is given by

$$J((X(1), C(1)), \ldots, (X(T), C(T))) = \sum_{t=1}^{T} Q(X(t), C(t)).$$

If we look at the controlled process in the $(\mathfrak{I}, \mathcal{E}, \chi)$ framework we see that the law of motion f determines the environment of the adaptive system. A problem in control becomes a problem of adaptation when there is significant uncertainty about the law of motion f; that is, it is only known that $f \in \{f_E, E \in \mathcal{E}\}$. Such problems are generally unsolvable by contemporary methods of optimal control theory (cf., for example, the comments of Tsypkin [1971, p. 178]). Clearly under such circumstances the adaptive plan will have to try out various policies in an attempt to determine a good one. To fix ideas, let us assume that each policy ${}^1A \in \mathcal{A}_1$ can be assigned an average or expected performance $\bar{Q}({}^1A, f)$ for each possible f. Moreover let us assume that this average can be estimated as closely as desired by simply trying ${}^1A$ long enough from any arbitrary time t onward. The object then is to search for the policy in $\mathcal{A}_1$ with the best average performance $\bar{Q}$, exploiting the best among *known* possibilities at each step along the way.

A control policy ${}^1A \in \mathcal{A}_1$ generates a sequence of control parameters $\langle C(t)\rangle$. Different trials of the policy ${}^1A$, say at times $t_1, t_2, \ldots, t_k$, will in general elicit different costs $Q(t_1), Q(t_2), \ldots, Q(t_k)$. However, the $(\mathfrak{I}, \mathcal{E}, \chi)$ framework requires that each $A \in \mathcal{A}$ be assigned a unique cost $\mu_E(A)$. To satisfy this requirement we can let $\mathcal{A} = \mathcal{A}_1 \times \mathfrak{N}$ where $\mathfrak{N}$ is the set of natural numbers $\{1, 2, 3, \ldots\}$. Then unique elements of $\mathcal{A}$, namely $({}^1A, t_1), ({}^1A, t_2), \ldots, ({}^1A, t_k)$, correspond to the successive trials of ${}^1A$ and the cost $Q(t_i)$ of trial t_i can be assigned as required,

$$\mu_E(A) = \mu_E(({}^1A, t_i)) = Q(t_i).$$

An adaptive plan τ will modify the policy at intervals on the basis of observed costs. With the definition of $\mathcal{A}$ just given this means that, if ${}^1A$ is tried at time t and is to be retained for trial at time $t + 1$,

$$\tau(I(t), \mathcal{A}(t)) = \tau(I(t), ({}^1A, t)) = ({}^1A, t + 1);$$

on the other hand, if a new policy ${}^1A'$ is to be tried,

$$\tau(I(t), \mathcal{A}(t)) = \tau(I(t), ({}^1A, t)) = ({}^1A', t + 1).$$

A sophisticated adaptive plan will probably retain a measure of the average performance of various policies tried so that $\mathcal{A}$ would be further extended by a component $\mathfrak{M}$ (see section 2.2) to $\mathcal{A} = \mathcal{A}_1 \times \mathfrak{N} \times \mathfrak{M}$. A still more sophisticated plan will progressively reduce uncertainty about the environment by deliberately selecting elements of C to elicit critical information, perhaps constructing a model of f_E. Then by exploiting predictions of the model τ can adjust the sequence $\langle C(t) \rangle$ to better performance as measured by the function J. At this level the illustration concerning searches, pattern recognition, and statistical inference applies in toto. If the plan is to be a payoff-only plan, then

$$I(t) = \mu_E(\mathcal{A}(t)) = Q(t),$$

and $\mathfrak{M}(t + 1)$ is updated by using $Q(t)$ in a recalculation of the average performance of $\mathcal{A}_1(t)$.

Finally the function J determines a ranking for every control sequence $\langle C(t) \rangle$, whether or not it is generated by a single policy. That is, an adaptive plan τ confronted with a law of motion f_E may try several policies, thereby generating a control sequence which no *single* ${}^1A \in \mathcal{A}_1$ could generate. However every control action $C(t)$ has a definite cost $Q(t)$. Thus the trajectory $\langle C(t) \rangle$ through C generated by τ can be ranked according to J. In this way J determines a criterion for ranking any $\tau \in \mathfrak{I}$ in any $E \in \mathcal{E}$. As a specific example, consider the case where the object is minimization of cumulative error. By assigning maximum payoff to the target region and reducing the payoff of other states in proportion to the associated error, the performance of a plan τ can be measured in terms of the cumulative payoff function $U_E(\tau, t)$. The greater $U_E(\tau, t)$ the less the cumulative error to time t.

The foregoing discussion can be made applicable to function optimization by so arranging it that a single "trial" of a policy ${}^1A$ produces a sufficient estimate of its average performance $\bar{Q}({}^1A,f)$. For instance the policy could be repeatedly tried over some extended interval of time which would then be taken to be a single time-step in the discrete formulation. In any case let us assume that each ${}^1A \in \mathcal{A}_1$ has a unique value $\mu_E({}^1A) = \bar{Q}({}^1A,f)$ which can be determined (with sufficient accuracy) in a single time-step. Moving the problem of estimating $\bar{Q}$ into the background in this way, reduces the control objective to finding the optimum of the function μ_E.

If the elements of $\mathcal{A}_1$ are represented as points in an n-dimensional Euclidean space $\mathcal{R}^n$ the problem becomes one of optimizing an n-dimensional (nonlinear) real function. For example the elements of $\mathcal{A}_1$ can be represented as strings of length n over some basic alphabet Σ. Since Σ can be recoded as a subset of $\{0, 1\}^m$, where m is the first integer greater than $\log_2$ (card Σ), this can be looked upon as optimization of an n-dimensional function having m-place binary fractions as arguments. (Thus, if $\Sigma = \{\sigma_0, \sigma_1, \sigma_2, \ldots\}$, the coding $\sigma_0 \leftrightarrow .00\ldots00$, $\sigma_1 \leftrightarrow .00\ldots01$, $\sigma_2 \leftrightarrow .00\ldots10$, etc., can be used. Then with ${}^1A \in \mathcal{A}_1$ represented by the string $\sigma_2\sigma_2\sigma_1$, say, the argument of μ_E becomes $(.00\ldots010, .00\ldots010, .00\ldots001)$.)

With this arrangement an adaptive plan τ uses its operators to generate a sequence of points $\mathcal{A}_1(1), \mathcal{A}_1(2), \mathcal{A}_1(3), \ldots$ converging to an optimum, much in the manner of standard iterative procedures. The adaptive approach, however, suggests important differences in what information from prior calculations should be retained (in $\mathcal{M}(t)$) in preparation for generation of the next point $\mathcal{A}_1(t+1)$. In particular certain adaptive plans proceed simultaneously and efficiently with global and local optimization of μ_E. (See chapters 4 and 5 for basic techniques.)

In the case of function optimization, high-dimensionality and nonlinearity of the function to be optimized (μ_E), in all but a few special cases, constitute insurmountable barriers to standard optimization algorithms. In the general control problem there is the added difficulty of nonstationarity. The schemata concept (first interpreted in function optimization terms in chapter 4, pp. 70–71) and the algorithms based upon it (chapter 6) provide specific remedies for the first two problems. The latter problem is substantial and difficult—it is discussed in section 9.3.

Summarizing:

$\mathcal{A}$, [control] a set having as its basic component the set of admissible control policies $\mathcal{A}_1$ augmented by a memory component $\mathcal{M}$ and a set of time subscripts $\mathcal{N}$ so that $\mathcal{A}(t) = (\mathcal{A}_1(t), t, \mathcal{M}(t))$; [function optimization] the domain of the function $\bar{Q}$ to be optimized.

Ω, [control] procedures for generating a new control policy from some set of given policies; [function optimization] procedures for generating a new point in the domain of $\bar{Q}$ from some set of given points.

$\mathfrak{J}$, plans for applying procedures from Ω to generate new policies [control] or points [function optimization] on the basis of observations.

$\mathcal{E}$, an indexing set corresponding to the initial uncertainty about the "law of motion" or the function to be optimized.

χ, the extension of the ranking J on control sequences to the plans inducing the sequences, assigning $\tau \in \mathfrak{J}$ the average (or minimum, etc.) ranking over the uncertainty indexed by $\mathcal{E}$.

6. CENTRAL NERVOUS SYSTEMS

> Behavior is primarily adaptation to the environment under sensory guidance. It takes the organism away from harmful events and toward favorable ones, or introduces changes in the immediate environment that make survival more likely.
>
> Hebb in *A Textbook of Psychology* (pp. 44–45)

I introduce this last example of an adaptive system with some hesitation. Not because the central nervous system (CNS hereafter) lacks qualifications as an adaptive system—on the contrary, this complex system exhibits a combination of breadth, flexibility, and rapidity of response unmatched by any other system known to man—but because there is so little prior mathematical theory aimed at explaining adaptive aspects of the CNS. Even an intuitive understanding of the relation between physiological micro-data and behavioral macro-data is only sporadically available. Perforce, mathematical theories enabling us to see some overall action of the CNS as a consequence of the actions and interactions of its parts are, when available at all, in their earliest formative stages.

Here, more than with the other examples, the initial advantage of the formal framework will be restatement of the familiar in a broader context. The best that can be hoped for at this stage is an occasional suggestion of new consequences of familiar facts: Without the advantages of a deductive theory, statements made within the framework can do little more than provide an experimenter with guideposts and cautions, suggesting possibilities and impossibilities, phenomena to anticipate, and conclusions to be accepted warily. This is a preliminary, heuristic stage marking the transition from unmathematical plausibility to the formal deductions of mathematical theory. In common with most heuristic and loose-textured arguments, it is difficult to eliminate ambiguities and contradictions—as with proverbs, proper application depends upon the intuition of the user. Specific applications of the formalism may arise, but these will probably be in areas of little uncertainty, where theory was not actually required; the formal procedures will be primarily corroborative. Only after considerable effort at this level can we hope for a theory mathematically rigorous and conceptually general—a genuinely predictive theory organizing large masses of data at many levels.

One of the earliest suggestions (or corroborations) from the formal underpinnings of the $(\mathfrak{J}, \mathfrak{E}, \chi)$ framework is quite fundamental. It can be established that major aspects of the behavior of any very complex system fall outside the explanatory power of simple input-output (*S-R* or switching) theory. This result is a rigorous version of the observation that ongoing activity in a complex system usually depends upon the past history of that system. This dependence, which both psychologists and computer theorists call "memory," finds its formal counterpart in the notion of state: distinct stimulus-state pairs generally giving rise to different responses. If there are many states (and, by any reasonable definition of state, the CNS has an astronomical number) the same stimulus may give rise to a great many different responses. Thus, observation of stimulus-response pairs will *not* enable us to discover the mode of operation of *any* system with a substantial number of states. For a system as complex as the CNS, such a result can be ignored only to the great detriment of the ensuing theory. It is a corollary of this result that complex systems can act in autonomous fashion, producing continuing response sequences in the absence of new stimulus. Thus, a stimulus may serve only to modify ongoing activity rather than to initiate it. In short, the responses of the CNS cannot be explained wholly in terms of concurrent stimuli.

The $(\mathfrak{J}, \mathfrak{E}, \chi)$ framework also emphasizes a second important point. An adequate theory must include more than a formal counterpart of the internal processes of the system being studied. The environment (or range of possible environments), the information received therefrom, and the ways the system can affect the environment, must also be represented. Moreover, the criterion χ emphasizes the importance of performance "along the way." The CNS cannot wait indefinitely for "useful" outcomes; some minimal level of ongoing performance is required. (E.g., if food is not obtained with sufficient frequency, death ensues, totally removing the possibility of further goal-oriented behavior.) Such observations are not new, but the $(\mathfrak{J}, \mathfrak{E}, \chi)$ framework does provide a form for fitting and arranging them, and it lends them emphasis. This at least gives us a fresh look at familiar facts, occasionally suggesting new consequences which might otherwise be overwhelmed in the plethora of macro- and micro-data (behavioral and physiological).

The discussion which follows will be based upon the informal theory of CNS action introduced by Hebb in his signal 1949 work and subsequently importantly enlarged by P. M. Milner (1957) and I. J. Good (1965). I will attempt a brief recapitulation of some of the main assumptions here, with the intention of orienting the reader having some knowledge of the area. This is only one view of CNS processes and the presentation has been kept deliberately simplistic. (E.g., a more sophisticated theory would take account of substantial evidence for distinct physiological mechanisms underlying short-term, medium-term, and long-term memory.) The object is to indicate, in as simple a context as possible, the relevance of the $(\mathfrak{I}, \mathcal{E}, \chi)$ framework to understanding the CNS as a means of "adaptation to the environment under sensory guidance." The reader without a relevant background can gain a significant understanding by reading the papers of Milner and Good; a reading of Hebb's excellent textbook (1958) will give a much more comprehensive view.

The basic element of Hebb's theory is the *cell assembly.* It is assumed to exhibit the following essential characteristics. (Comments in parentheses in the presentation of characteristics refer to possible neurophysiological mechanisms):

1. A cell assembly is formed in response to repetitions of some relatively simple critical feature of the sensory input, such as pressure on a particular skin area, a simple odor, a vowel sound, an increase of brightness, a line of particular slope in the visual field, and so on. (From a more physiological point of view, a cell assembly is taken to be a collection of hundreds of neurons interconnected via synapses of high conductivity.) After some assemblies have been formed others may be formed in response to the repetitive actions of already extant (precursor) assemblies.
2. Whenever the critical feature causing the cell assembly's formation subsequently appears in the sensory input, the cell assembly tends to become active. Cell assemblies formed in response to precursor assemblies tend to become active when their precursors are active. (Once a significant number of neurons in the cell assembly attain a high pulse rate, the remaining neurons quickly follow suit because of the highly conductive interconnections. Because the interconnections form many loops, a reverberation results and the neurons tend to remain active for a period of time long compared to the stimulus time.)
3. Cell assemblies exhibit changing positive and negative associations with each other. A cell assembly which is active increases the likelihood of activity in all assemblies with which it is positively associated and de-

creases the likelihood of activity in all assemblies with which it is negatively associated. Positive association between a pair of cell assemblies increases whenever they are active at the same time. Negative association is asymmetrical in that one cell assembly may be negatively associated with a second, while the second is not necessarily negatively associated with the first; this negative association increases each time the first assembly is active and the second is inactive. (The underlying neural assumption here is that, if neuron n_2 produces a pulse immediately after it receives a pulse from neuron n_1, then n_1 is better able to elicit a pulse from n_2 in the future; contrariwise, if n_2 produces no pulse upon receiving a pulse from n_1, then n_1 is more likely to inhibit n_2 in the future. It is usually assumed that this process is the result of changing synapse levels. The same process can be invoked in explaining the origin of cell assemblies.) It should be noted that, under this assumption, there is a tendency for cell assemblies to become active in fixed combinations, at the same time actively suppressing alternative combinations. (Because a cell assembly involves only a minute fraction of the neurons in a CNS, a great many can be excited at any instant, different configurations corresponding to different perceived objects, etc.) Temporal association (i.e., probable action sequences) can occur via appropriate asymmetries; e.g., assembly α can arouse β via positive association while β inhibits α through negative association. Thus the action sequence is always $\alpha\beta$, never the reverse.

4. At any instant the response of the CNS to sensory input is determined by the configuration of active cell assemblies. (Overt behavior such as eye movement, activation of reflexes, release of voluntary muscle sequences, and so on will accompany most sensory events. Via the mechanisms of (3), neurons involved in this behavior will tend to become components of cell assemblies active at the same time. Since pulse trains from the active cell assemblies dominate overall CNS activity, overt behavior will thus be determined by the active configurations. In effect, the sensory input modulates the ongoing activity in the CNS to produce overt behavior.)
5. Cell assemblies involved in temporal sequences yielding "need satisfaction" (satisfaction of hunger, thirst, etc.) have their associations enhanced; the greater the "need," the greater the enhancement. ("Needs" are internal conditions in the CNS-controlled organism, conditions primarily concerned with survival, which set basic restrictions on CNS

action. In typical environmental situations, overt behavior is required for "need satisfaction," and then the satisfaction is only temporary—the organism consumes the resources involved in order to maintain itself. Innate internal mechanisms in the organism automatically "reward" satisfaction of hunger, thirst, etc., and perhaps some more generalized needs such as curiosity. These "rewards" may be mediated by innately organized neural networks which exhibit increasing activity as a corresponding need increases. Such internally generated stimuli would progressively disturb established configurations and sequences, unless they resulted in reduction of the corresponding need. Ultimately, in the absence of satisfaction, this disruption would cause an increasingly broad search through the organism's behavioral repertory—a kind of hunt through increasingly unusual cell assembly configurations in an attempt to produce an appropriate overt response. Temporal associations of cell assemblies, active when such a disturbance is reduced, would retain their incremented synapse levels. Those active during a period of increasing disturbance would encounter subsequent interference, causing synapse level increments to be transitory. Assemblies having precursors occurring early in "need satisfaction" sequences acquire a particular role. They serve as "leading indicators," becoming active in advance of actual primitive needs; they may serve as "learned needs" [goals]. A hierarchy of precursors of precursors, etc., can provide the system with a hierarchy of "learned needs," some of them quite remote from the primitive needs. That is, assemblies containing substantial segments of the innately organized networks as components, or assemblies closely associated therewith, could give rise to secondary and higher-order "learned needs." The effects of these new assemblies will be much like those generated by the innately organized networks.)

6. An active cell assembly primes cell assemblies associated with it as successors in temporal sequences, making them more likely to be active subsequently. (A neuron producing pulses at a high rate tends to become fatigued, with a consequent drop in pulse rate. A neuron that is being inhibited tends to exhibit less fatigue than normal because of its very low pulse rate. A kind of inhibitory priming results, because the neuron is hyperresponsive once the inhibition ceases. It is also likely that priming occurs by transmission of "priming molecules" through the synapses of active neurons. Priming provides the CNS with expectations and predictions. In effect the system expects and is ready to respond to selected sets

of features from the myriads of possibilities. When primed cell assemblies subsequently become active—i.e., when the corresponding predictions are verified—the associations involved are strengthened by the mechanisms of (3). The resulting network of associations constitutes a model of the environment within the CNS. The model is dynamic in the sense that it takes sensory data as input and primes different temporal sequences on the basis of the model's predictions. Introspection confirms that, for the human CNS, models of the environment are indeed used to compare alternative courses of action. This model is ultimately dedicated to keeping primitive needs fulfilled, but it incorporates "leading indicators," etc., so that needs rarely become acute enough to determine action directly.)

How can the $(\mathfrak{I}, \mathcal{E}, \chi)$ framework help in analyzing this model? Certain analogies with other processes are suggestive. Individual cell assemblies act, in part, like the detectors in pattern recognizers: they are activated by particular features of the environment, features presumably relevant to the organism's needs. At the same time, the configuration of cell assemblies active at any given time defines the organism's response to the environment. In the terms used earlier, such a configuration is an element of the system's repertory. Assuming that the set of all cell assemblies is fixed (as it might be, to a first approximation, in a mature organism) or, at least, that the potentially available cell assemblies can be enumerated, the set of all possible assembly configurations constitutes the system's repertory of techniques for confronting the environment.

When cell assemblies are in mutual negative association (cross-inhibition), they act much as the alleles of a chromosomal locus; any active configuration can contain at most one of the assemblies, because it will actively suppress the others in the set. Positive associations between cell assemblies which favor particular configurations are analogous to the linkage of coadapted alleles in a chromosome. Indeed there are many potentially fruitful "genetic" analogies. As the CNS gains experience, some assemblies in a cross-inhibited set are likely to be expressed in a broadened range of environmental conditions, at the expense of others in the set—a process suggestive of the evolution of (partial) dominance. Various genetic operators such as crossover and inversion find their counterparts in the ways in which cell assembly associations are modified. Temporal associations correspond to feedback among gene-products and sequential expression of genes. The list can be extended easily.

The needs of the organism define its goals, and ultimately set a criterion on performance. Just as in games and searches, there is the problem that individual responses do not often directly yield need satisfaction. However, the internal model discussed in connection with property (6) enables the CNS to constantly improve performance in the absence of current need satisfaction. Two kinds of improvement are possible. First of all, cell assemblies typically respond to too broad a range of situations when first formed, yielding inconsistencies in the model. That is, situations activating the same combinations of cell assemblies, and hence the same responses, are followed by radically different outcomes. The remedy here is much like that for inadequacy of detectors discussed in the illustration on searches. Because of the inconsistencies new associations are formed between the cell assemblies involved, causing them to split and recombine so that their responses are more discriminative. (Hebb calls the related procedures fractionation and recruitment.) The second kind of improvement consists in "filling in" the model—generally there will be many situations where no expectations or predictions have been developed. This clearly provides an important role for curiosity. The CNS must experience a wide enough range of situations to provide an adequate repertory of *relevant* temporal sequences. Just as with the coadapted sets of genetics, the basic laws of cell assemblies permit flexible recombination (association) under environmental (sensory) guidance, the actual combinations being influenced by the parts of the model (associations) already extant. In this way a tremendous range of useful procedures can be formed from relatively few elements. More importantly, a single experience then constitutes a trial of a great many relevant associations, just as in genetics a single organism tests a great many coadapted sets. Property (3) assures that many associations will be tested and modified. The ultimate "survival" of various combinations of assemblies is determined by their consistency within the model and their success in contributing to learned or unlearned need satisfaction.

While the foregoing analogies are ready offshoots of the formal framework, the basic task of theory in this area is quite difficult. It must enable one to judge whether proposed mechanisms for CNS operation permit the learning rates, utilization of cues, transfer of learning, etc., that one actually observes. How does the CNS maintain its rapidity and appropriateness of response, while extending its breadth and filling in its model of the environment? Section 8.4 indicates one way in which concepts from the $(\mathfrak{I}, \mathcal{E}, \chi)$ framework can be brought to bear. In particular, the robustness of reproductive plans, when interpreted in this area, indicates some promising directions, but we even lack good general measures of performance here. A kind of error function based upon average need levels might be interesting for organisms not quite so efficient as man at keeping their primitive needs satisfied. A criterion χ could then be formulated, much as it was in the optimal control illustration, and it might be possible to use the framework more precisely in this context (especially for animals in the wild state). Some suggestions for bringing cell assembly theory within the range of the $(\mathfrak{T}, \mathcal{E}, \chi)$ framework are made in section 8.4. There is much to be done before we can hope for definite, general results from theory.

Summarizing:

$\mathfrak{A}$, repertory of possible cell assemblies.

Ω, possible association rules (Hebb's rule for synapse change, short-term memory rules, etc.).

$\mathfrak{T}$, possible (or hypothetical) organizations of the CNS in terms of conditions under which the rules of Ω are to operate.

$\mathcal{E}$, the range of environments in which the CNS being studied is expected to operate (relevant features, cues, etc.).

χ, the ranking of organizations in $\mathfrak{T}$ according to performance over $\mathcal{E}$, for example, according to ability to keep average needs low under any situation in $\mathcal{E}$ (cf. optimal control illustration).

These illustrations are intended to demonstrate the broad applicability of the $(\mathfrak{T}, \mathcal{E}, \chi)$ framework. They can also serve to demonstrate something else. The obstacles described informally at the end of section 1.2 do indeed appear as central problems in each of the fields examined. This is an additional augury for making a unified approach to adaptation—common problems should have common solutions. Much of the work that follows is directed to the resolution of these general problems. In section 9.1 the problems are listed again, more formally, and the relevance of this work to their resolution is recapitulated.

4. Schemata

An adaptive system faces its principal challenge when the set of possible structures $\mathcal{A}$ is very large and the performance functions μ_E involve many local maxima. It is important then for the adaptive system to provide itself with whatever insurance it can against a prolonged search. It is clear that the search of $\mathcal{A}$ must go on so long as significant improvements are possible (unless the system is to settle for inferior performance throughout the remainder of its history). At the same time, unless it exploits possibilities for improved performance *while* the search goes on, the system pays the implicit cost of a performance less even than the best among *known* alternatives. Moreover, unexploited possibilities may contain the key to optimal performance, dooming the system to fruitless search until they are implemented. There is only one insurance against these contingencies. The adaptive system must, as an integral part of its search of $\mathcal{A}$, persistently test and incorporate structural properties associated with better performance. As with most insurance, this particular policy contains a limiting clause: useful properties must be identified to be exploited. The present chapter is concerned with this limitation.

Almost by definition useful properties are points of comparison between structures yielding better-than-average performance. The question then is: How are the structures in $\mathcal{A}$ to be compared? If the structures are built up from components, comparison in terms of common components is natural and the question becomes: How is credit for the above-average performance of a structure to be apportioned to its components? A more general approach uses feature detectors (see section 3.4) to make comparisons. Since one can find an appropriate detector for any effectively describable feature of structures in $\mathcal{A}$ (including the presence or absence of given components) this approach is well suited to present purposes.

To begin with let us see how comparisons can be developed when a finite set of detectors $\{\delta_i : \mathcal{A} \to V_i, i = 1, \ldots, l\}$ is given. In terms of the given detectors each structure $A \in \mathcal{A}$ will have a *representation* $(\delta_1(A), \delta_2(A), \ldots, \delta_l(A))$; that is, each structure A will be described by its particular ordered set of l *attributes* or detector values $\delta_i(A) \in V_i$, $i = 1, \ldots, l$. Thus, for a chromosome A, V_i can designate the set of alleles of locus i (see section 3.1) and the corresponding representation of A is the specification of the ordered set of alleles which make up the chromosome. For a von Neumann economy (section 3.2) the V_i can designate the possible levels of the ith activity so that the representation of a mixture of activities is simply the corresponding activity vector. Similar considerations apply to each of the remaining illustrations of chapter 3. Clearly, with a given set of l detectors, two structures will be distinguishable only insofar as they have distinct representations. Since, in the present chapter, we are only interested in comparisons let us assume that all structures in $\mathcal{A}$ are distinguishable (have distinct representations) or, equivalently, that $\mathcal{A}$ is used to designate distinguishable subsets of the original set of structures. For simplicity in what follows $\mathcal{A}$ will simply be taken to *be* the set of representations provided by the detectors (rather than the abstract elements so represented).

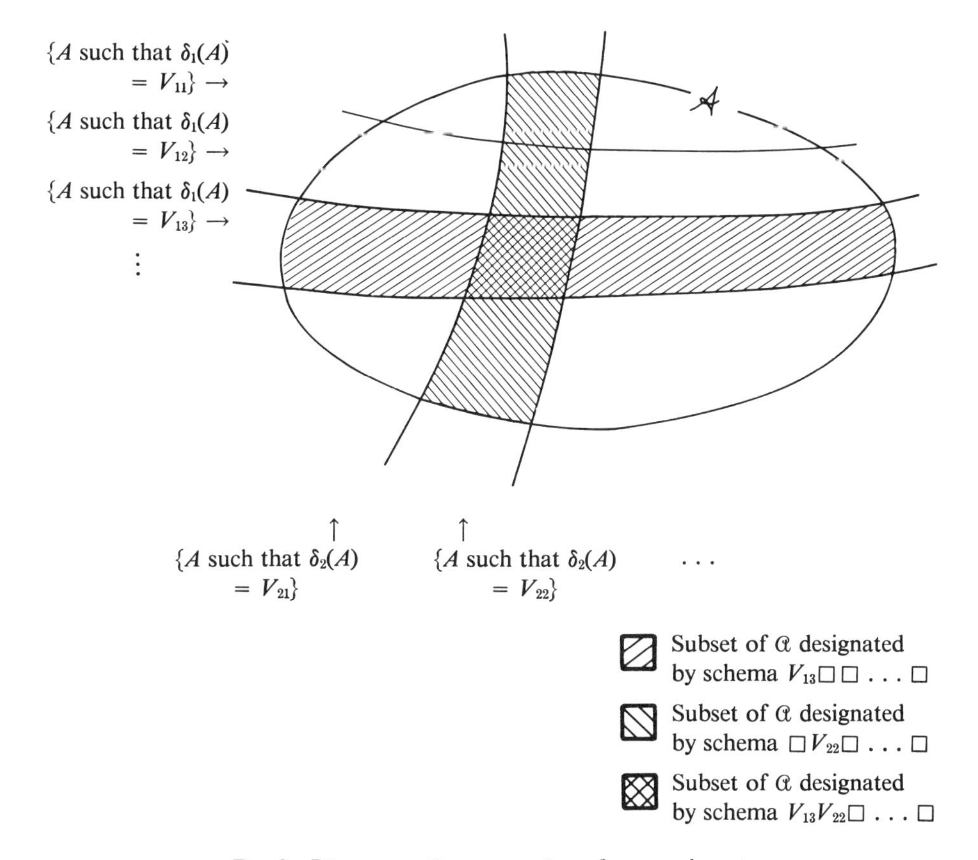

Fig. 9. Diagrammatic presentation of some schemata

Now our objective is to designate subsets of $\mathcal{A}$ which have attributes in common. To do this let the symbol "$\square$" indicate that we "don't care" what attribute occurs at a given position (i.e., for a given detector). Thus $(v_{13}, \square, \square, \ldots, \square)$ designates the subset of all elements in $\mathcal{A}$ having the attribute $v_{13} \in V_1$. (Equivalently, $(v_{13}, \square, \ldots, \square)$ designates the set of all l-tuples in $\mathcal{A}$ beginning with the symbol v_{13}; hence, for $l = 3$, (v_{13}, v_{22}, v_{32}) and (v_{13}, v_{21}, v_{31}) belong to $(v_{13}, \square, \square)$, but (v_{12}, v_{22}, v_{32}) does not.) The set of all l-tuples involving combinations of "don't cares" and attributes is given by the augmented product set $\Xi = \Pi_{i=1}^{l} \{V_i \cup \{\square\}\}$. Then any l-tuple $\xi = (\Delta_{i_1}, \Delta_{i_2}, \ldots, \Delta_{i_l}) \in \Xi$ designates a subset of $\mathcal{A}$ as follows: $A \in \mathcal{A}$ belongs to the subset if and only if (i) whenever $\Delta_{i_j} = \square$, any attribute from V_j may occur at the jth position of A, and (ii) whenever $\Delta_{i_j} \in V_j$, the attribute Δ_{i_j} must occur at the jth position of A. (For example, $(v_{11}, v_{21}, v_{31}, v_{43})$ and $(v_{13}, v_{21}, v_{32}, v_{43})$ belong to $(\square, v_{21}, \square, v_{43})$ but $(v_{11}, v_{21}, v_{31}, v_{42})$ does not.) The set of l-tuples belonging to Ξ will be called the set of *schemata;* Ξ amounts to a decomposition of $\mathcal{A}$ into a large number of subsets based on the representation in terms of the l detectors $\{\delta_i : \mathcal{A} \rightarrow V_i, i = 1, \ldots, l\}$.

Schemata provide a basis for associating combinations of attributes with potential for improving current performance. To see this, let "improvement" be defined as any increment in the average performance over past history. That is, if $\mu(\mathcal{A}(t))$ is the performance of the structure $\mathcal{A}(t)$ tried at time t, the object is to discover ways of incrementing

$$\bar{\mu}(T) = \frac{1}{T} \sum_{t=1}^{T} \mu(\mathcal{A}(t)).$$

(A more sophisticated measure would give more weight to recent history, using

$$\mu'(T) = (\textstyle\sum_{t=1}^{T} c_t \mu(\mathcal{A}(t)))/(\sum_{t=1}^{T} c_t),\ c_t > c_{t'} \quad \text{for } t > t',$$

but the simple average suffices for the present discussion.) Though $\bar{\mu}(T)$ can be incremented by simply repeating the structure yielding the best performance up to time T this does not yield new information. Hence the object is to find *new* structures which have a high probability of incrementing $\bar{\mu}(T)$ significantly. An adaptive plan can use schemata to this end as follows: Let $A \in \mathcal{A}$ have a probability $P(A)$ of being tried by the plan τ at time $T + 1$. That is, τ induces a probability distribution P over $\mathcal{A}$ and, under this distribution, $\mathcal{A}$ becomes a sample space. The performance measure μ then becomes a random variable over $\mathcal{A}$, $A \in \mathcal{A}$ being tried with probability $P(A)$ and yielding payoff $\mu(A)$. More importantly, any schema $\xi \in \Xi$ designates an event on the sample space $\mathcal{A}$. Thus, the restriction $\mu \mid \xi$ of μ to the subset designated by ξ, is also a random variable, $A \in \xi$ being tried with probability $(P(A))/(\sum_{A' \in \xi} P(A'))$ and yielding payoff $\mu(A)$. In what follows ξ will be used to designate both an element of Ξ and the corresponding random variable with sample space ξ, the particular usage being specified when it is not clear from context. As a random variable, ξ has a well-defined average μ_ξ (and variance σ_ξ^2) where, intuitively, μ_ξ is the payoff expected when an element of ξ is randomly selected under the marginal distribution $(P(A))/(\sum_{A' \in \xi} P(A'))$.

Clearly, when $\mu_\xi > \bar{\mu}(T)$, instances of ξ (i.e., $A \in \xi$) are likely to exhibit performance better than the current average $\bar{\mu}(T)$. This suggests a simple procedure (a bit too simple as it turns out) for exploiting combinations of attributes associated with better-than-average performance while further searching $\mathcal{A}$: (i) try instances of various schemata until at least one schema ξ is located which exhibits a sample average $\hat{\mu}_\xi > \bar{\mu}(T)$; (ii) generate *new* instances of the (observed) above-average schema ξ, returning to step (i) from time to time (particularly when the increasing overall average $\bar{\mu}(T)$ comes close to $\hat{\mu}_\xi$) to locate new schemata $\{\xi'\}$ for which $\hat{\mu}_{\xi'} > \bar{\mu}(T)$. In effect, then, credit is apportioned to a combination of attributes in accord with the observed average performance of its instances. This procedure has some immediate advantages over a fixed random (or enumerative) search of $\mathcal{A}$: it generates improvements with high probability while gathering new information by trying new $A \in \mathcal{A}$; furthermore, the new trials of the above-average schema ξ increase confidence that the observed average $\hat{\mu}_\xi$ closely approximates μ_ξ. It is oversimple because each instance $A \in \mathcal{A}$ tried yields information about a great many schemata other than ξ—information which is not used.

Given l detectors, a single structure $A \in \mathcal{A}$ is an instance of 2^l distinct schemata, as can be easily affirmed by noting that A is an instance of any schema ξ defined by substituting "□"s for one or more of the l attribute values in A's representation. Thus a single trial A constitutes a trial of 2^l distinct random variables, yielding information about the expected payoff μ_ξ of each. (If l is only 20 this is still information about a million schemata!) Any procedure which uses even a fraction of this information to locate ξ for which $\mu_\xi > \bar{\mu}(T)$ has a substantial advantage over the one-at-a-time procedure just proposed.

Exploiting this tremendous flow of information poses a much more clearly defined challenge than the one which started the chapter. Schemata have advanced our understanding, in this sense, but the new problem is difficult. The amount of storage required quickly exceeds all feasible bounds if one attempts to record the average payoff of the observed instances of each schema sampled. Moreover, the information will be employed effectively only if it is used to generate new $A \in \mathcal{A}$ which, individually, test as many above-average schema as possible. The adaptive system is thus faced with a specific problem of compact storage, access, and effective use of information about extremely large numbers of schemata. Chapter 6 ("Reproductive Plans and Genetic Operators") sets forth a resolution of these difficulties, but a closer look at schemata (the remainder of this chapter) and the optimal allocation of trials to sets of schemata (the next chapter) provides the proper setting.

Let us begin with a concrete, but fairly general, interpretation of schemata stemming from the earlier discussion of control and function optimization (section 3.5, p. 57). Consider an arbitrary bounded function $f(x)$, $0 \leqq x < 1$, and assume that x is specified to an accuracy of one part in a million or, equivalently, that values of x are discretely represented by 20 bits. Define $\mathcal{A}$ to be the set of 2^{20} discrete values of x represented with 20 detectors $\{\delta_j: \mathcal{A} \rightarrow \{0, 1\}, j = 1, \ldots, 20\}$ where $\delta_j(x)$, $x \in \mathcal{A}$, assigns to x the value of the jth bit in the binary expansion of x. The schema $1\square\square \ldots \square$ then is just the right half-plane $\frac{1}{2} \leqq x < 1$, while the schema $\square\square 0 \square \ldots \square$ is a set of four strips $\{0 \leqq x < \frac{1}{8}, \frac{1}{4} \leqq x < \frac{3}{8}, \frac{1}{2} \leqq x < \frac{5}{8}, \frac{3}{4} \leqq x < \frac{7}{8}\}$ and the schema $1\square 0 \square \ldots \square$ is the intersection of the two previous schemata $\{\frac{1}{2} \leqq x < \frac{5}{8}, \frac{3}{4} \leqq x < \frac{7}{8}\}$ (see Figure 10).

With this representation there are 3^{20} distinct schemata since any 20-tuple over the set $\{0, 1, \square\}$ defines a schema. (More technically, the schemata are simply hyperplanes, of dimension 20 or less, in the 20-dimensional space of detector-value combinations.) Note that there are many points, such as $x = \frac{13}{16} = .11010 \ldots 0$, which are instances of all three of the schemata just singled out. Note also that f has a well-defined average value f_ξ on each schema ξ (for any weighting of the values $f(x)$, as by a probability distribution). Clearly, for any x, knowledge of $f(x)$ is relevant to estimating f_ξ for any schema for which $x \in \xi$. Moreover, observations

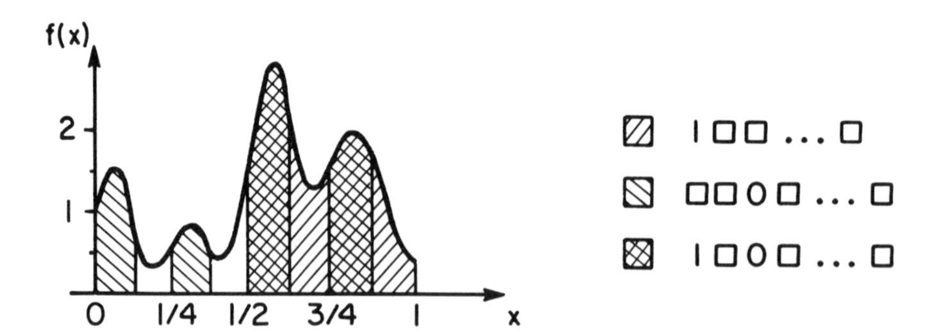

Fig. 10. Some schemata for a one-dimensional function

of f for relatively few x will enable f_ξ to be estimated for a great many $\xi \in \Xi$. Even a sequence of four observations, say $x(1) = .0100010 \ldots 0$, $x(2) = .110100 \ldots 0$, $x(3) = .100010 \ldots 0$, $x(4) = .1111010 \ldots 0$, enables one to calculate three-point estimates for many schemata, e.g. (assuming all points are equally likely or equally weighted), $\hat{f}_{1\square\square\ldots\square} = (f(x(2)) + f(x(3)) + f(x(4)))/3$ and $\hat{f}_{\square\square\square\square 01\square\ldots\square} = (f(x(1)) + f(x(3)) + f(x(4)))/3$, and two-point estimates for even more schemata, e.g., $\hat{f}_{\square 1\square\square 01\square\ldots\square} = (f(x(3)) + f(x(4)))/2$ and $\hat{f}_{11\square\ldots\square} = (f(x(2)) + f(x(4)))/2$.

The picture is not much changed if f is a function of many variables $x_1, \ldots, x_d$. Using binary representations again, we now have $20d$ detectors (assuming the same accuracy as before), 3^{20d} schemata, and each point is an instance of 2^{20d} schemata. In the one-dimensional case the representation transformed the problem to one of sampling in a 20-dimensional space—already a space of high dimensionality—so the increase to a $20d$-dimensional space really involves no significant conceptual changes. Interestingly, each point $(x_1, \ldots, x_d)$ is now an instance of 2^{20d} schemata rather than 2^{20} schemata, an exponential (dth power) increase. Thus, for a given number of points tried, we can expect an exponential (dth power) increase in the number of schemata for which f_ξ can be estimated with a given confidence. As a consequence, if the information about the schemata can be stored and used to generate relevant new trials, high dimensionality of the argument space $\{0 \leqq x_j < 1, j = 1, \ldots, d\}$ imposes no particular barrier.

It is also interesting in this context to compare two different representations for the same underlying space. Six detectors with a range of 10 values can yield approximately the same number of distinct representations as 20 detectors with a range of 2 values, since $10^6 \cong 2^{20} = 1.05 \times 10^6$ (cf. decimal encoding vs. binary encoding). However the numbers of schemata in the two cases are vastly different: $11^6 = 1.77 \times 10^6$ vs. $3^{20} = 3.48 \times 10^9$. Moreover in the first case each $A \in \mathcal{A}$ is an instance of only $2^6 = 64$ schemata, whereas in the second case each $A \in \mathcal{A}$ is an instance of $2^{20} = 1.05 \times 10^6$ schemata. This suggests that, for adaptive plans which can use the increased information flow (such as the reproductive plans), many detectors deciding among few attributes are preferable to few detectors with a range of many attributes. In genetics this would correspond to chromosomes with many loci and few alleles per locus (the usual case) rather than few loci and many alleles per locus.

Returning to the view of schemata as random variables, it is instructive to determine how many schemata receive at least some given number $n < N$ of trials when N elements of $\mathcal{A}$ are selected at random. This will give us a better idea of the *intrinsic parallelism* wherein a sequence of trials drawn from $\mathcal{A}$ is at the same time a (usually shorter) sequence of trials for each of a large number of schemata

$\xi \in \Xi$. It will be helpful in approaching this calculation (and in later discussions) to more carefully classify the schemata. A schema will be said to be *defined on* the set of positions $\{i_1, \ldots, i_h\}$ at which $\Delta_{i_j} \neq \square$. If $V = \bigcup_i V_i$ has k elements and we consider all schemata over V, i.e., $\Xi = \{V \cup \{\square\}\}^l$, then there are k^h distinct schemata defined on any given set of $h \leqq l$ positions. Moreover, for any given set of h positions, *every* $A \in V^l$ is an instance of one of these k^h schemata. That is, the set of schemata defined on a given set of positions partitions $\mathcal{A}$, and each distinct set of positions gives rise to a different partition of $\mathcal{A}$. (For example, if $V = \{0, 1\}$ and $l = 4$, the set of schemata defined on position 1 is $\{0\square\square\square, 1\square\square\square\}$, where $0\square\square\square$ abbreviates $(0, \square, \square, \square)$ etc. It is clear that every element in $\mathcal{A} = V^l$ begins either with the symbol 0 or else the symbol 1, hence the given set partitions $\mathcal{A}$. Similarly the set defined on position 2, $\{\square 0\square\square, \square 1\square\square\}$, partitions $\mathcal{A}$, and the set defined on positions 2 and 4, $\{\square 0\square 0, \square 0\square 1, \square 1\square 0, \square 1\square 1\}$ is still a different partition of $\mathcal{A}$, a refinement of the one just previous.) There are $\binom{l}{h}$ distinct ways of choosing h positions $\{1 \leqq i_1 < i_2 < \cdots < i_h \leqq l\}$ along the l-tuple, and h can be any number between 1 and l. Thus there are $\sum_{h=1}^{l} \binom{l}{h} = 2^l - 1$ distinct partitions induced on $\mathcal{A}$ by these sets of schemata. It follows that any sequence of N trials of $\mathcal{A}$ will be simultaneously distributed over each of these partitions. That is, *each* of the 2^l *sets* of schemata defined on the 2^l distinct choices of positions receives N trials.

On the assumption that elements of $\mathcal{A}$ are tried at random, uniformly (elements equally likely) and independently, we can use the Poisson distribution to determine the number of schemata receiving at least $n < N$ trials. The basic parameter required is the average number of trials per schema for any set of schemata defined on h positions. The value of this parameter is just N/k^h since there are k^h schemata defined on a fixed set of h positions. The Poisson distribution then gives

$$r(n, N) = \sum_{n'=n}^{\infty} ((N/k^h)^{n'}/n'!) \exp(-N/k^h)$$

as the proportion of schemata defined on the positions $i_1, \ldots, i_h$ and receiving at least n out of the N trials.

This can be directly generalized to give a lower bound in the case where the distribution over $\mathcal{A}$ is no longer uniform. Let $\chi_\epsilon(i_1, \ldots, i_h)$ designate the fraction of schemata defined on $i_1, \ldots, i_h$ for which the probability of a trial is at least ϵ/k^h, let γ_h be the proportion of the $\binom{l}{h}$ sets of schemata defined on h positions for which $\chi_\epsilon(i_1, \ldots, i_h) > \beta_0$ and, finally, let $\gamma_0 = \min_h \gamma_h$. Then the expression above, by a simple manipulation, yields

$$\begin{aligned} r(n, \epsilon, N) &= \sum_{h=1}^{l} \gamma_h \binom{l}{h} \beta_0 k^h \sum_{n'=n}^{\infty} ((\epsilon N/k^h)^{n'}/n'!) \exp(-\epsilon N/k^h) \\ &\geqq \gamma_0 \beta_0 \sum_{h=1}^{l} \binom{l}{h} k^h \sum_{n'=n}^{\infty} ((\epsilon N/k^h)^{n'}/n'!) \exp(-\epsilon N/k^h). \end{aligned}$$

5. The Optimal Allocation of Trials

In the last chapter a schema ξ was defined as potentially useful when $\hat{\mu}_\xi$, the observed average performance of instances of that schema, was significantly greater than the overall average performance. However, $\hat{\mu}_\xi$ is basically a sample average for a random variable (or sequence of random variables) and, as such, is subject to sampling error. For any two schemata ξ and ξ', there is always a non-zero probability that $\mu_{\xi'} > \mu_\xi$ even though $\hat{\mu}_\xi > \hat{\mu}_{\xi'}$. This reintroduces in a sharp form the conflict of exploiting what is known vs. obtaining new information. Confidence that the ranking $\hat{\mu}_\xi > \hat{\mu}_{\xi'}$ reflects a true ranking $\mu_\xi > \mu_{\xi'}$ can be increased significantly only by allocating additional trials to *both* ξ and ξ'. Thus, we can allocate a trial to exploit the observed best or we can allocate a trial to reduce the probability of error as much as possible but we cannot generally do both at once. Given a string-represented domain $\mathcal{A}$, it is important to have some idea of what proportion of trials should be allocated to each purpose as the number of trials increases.

Corresponding to each of these objectives—exploitation vs. new information—there is a source of loss. A trial allocated to the *observed* best may actually incur a loss because of sampling error, the observed best being in fact less than the best among the alternatives examined. On the other hand trials intended to maximally reduce the probability of error will generally be allocated to a schema other than the observed best. This means a performance less on the average than the best among known alternatives, when the observations reflect the true ranking. Stated succinctly, more information means a performance loss, while exploitation of the observed best runs the risk of error perpetuated.

Competing sources of loss suggest the possibility of optimization—minimizing expected losses by a proper allocation of trials. If we can determine the optimal allocation for arbitrary numbers of trials, then we can determine the minimum losses to be expected as a function of the number of trials. This in turn can be used as a criterion against which to measure the performance of suggested adaptive plans. Such a criterion will be particularly useful in determining the worth of plans which use schemata to compare structures in $\mathcal{A}$. The objective of this chapter is to determine this criterion. In the process we will learn a good deal more about schemata and intrinsic parallelism.

1. THE 2-ARMED BANDIT

The simplest precise version of the optimal allocation problem arises when we restrict attention to two random variables, ξ and ξ', with only two possible payoffs, 0 or 1. A trial of ξ produces the payoff 1 with probability p_1 and the payoff 0 with probability $1 - p_1$; similarly ξ' produces 1 with probability p_2 and 0 with probability $1 - p_2$. (For example, such trials could be produced by flipping either of two unbalanced coins, one of which produces heads with probability p_1, the other with probability p_2.) One is allowed N trials on each of which either ξ or ξ' can be selected. The object is to maximize the total payoff (the cumulative number of heads). Clearly if we know that ξ produces payoff 1 with a higher probability then all N trials should be allocated to ξ with a resulting expected accumulation $p_1 \cdot N$. On the other hand if we know nothing initially about ξ and ξ' it would be unwise not to test both. How trials should be allocated to accomplish this is certainly not immediately obvious. (This is a version of the much-studied 2-armed bandit problem, a prototype of important decision problems. Bellman [1961] and Hellman and Cover [1970] give interesting discussions of the problem.)

If we allow the two random variables to be completely general (having probability distributions over an arbitrary number of outcomes), we get a slight generalization of the original problem which makes direct contact with our discussion of schemata. The outcome of a trial of either random variable is to be interpreted as a payoff (performance). The object once more is to discover a procedure for distributing an arbitrary number of trials, N, between ξ and ξ' so as to maximize the expected payoff over the N trials. As before, if we know for each ξ_i the mean and variance (μ_i, σ_i^2) of its distribution (actually the mean μ_i would suffice), the problem has a trivial solution (allocate all trials to the random variable with maximal mean). The conflict asserts itself, however, if we inject just a bit more uncertainty. Thus we can know the mean-variance pairs but not which variable is described by which pair; i.e., we know the pairs (μ_1, σ_1^2) and (μ_2, σ_2^2) but not which pair describes ξ.

If it could be determined through some small number of trials which of ξ and ξ' has the higher mean, then from that point on all trials could be allocated to that random variable. Unfortunately, unless the distributions are non-overlapping, no finite number of observations will establish *with certainty* which random variable has the higher mean. (E.g., given $\mu_\xi > \mu_{\xi'}$ along with a probability $p > 0$ that a trial of ξ' will yield an outcome $x > \mu_\xi$, there is still a probability p^N after N

trials of ξ' that *all* of the trials have had outcomes exceeding μ_ξ. A fortiori their average $\hat{\mu}_{\xi'}$ will exceed μ_ξ with probability at least p^N, even though $\mu_{\xi'} < \mu_\xi$.) Here the tradeoff between gathering information and exploiting it appears in its simplest terms. To see it in exact form let $\xi_{(1)}(N)$ name the random variable with the highest *observed* payoff rate (average per trial) after N trials and let $\xi_{(2)}(N)$ name the other random variable. For any number of trials n, $0 \leqq n \leqq N$, allocated to $\xi_{(2)}(N)$ (and assuming overlapping distributions) there is a positive probability, $q(N - n, n)$, that $\xi_{(2)}(N)$ is actually the random variable with the highest mean, max $\{\mu_\xi, \mu_{\xi'}\}$. The two possible sources of loss are: (1) The *observed* best $\xi_{(1)}(N)$ is really second best, whence the $N - n$ trials given $\xi_{(1)}(N)$ incur an (expected) cumulative loss $(N - n)\cdot|\mu_\xi - \mu_{\xi'}|$; this occurs with probability $q(N - n, n)$. (2) The *observed* best is in fact the best, whence the n trials given $\xi_{(2)}(N)$ incur a loss $n\cdot|\mu_\xi - \mu_{\xi'}|$; this occurs with probability $(1 - q(N - n, n))$. The total expected loss for any allocation of n trials to $\xi_{(2)}$ and $N - n$ trials to $\xi_{(1)}$ is thus

$$L(N - n, n) = [q(N - n, n)\cdot(N - n) + (1 - q(N - n, n))\cdot n]\cdot|\mu_\xi - \mu_{\xi'}|.$$

We shall soon see that, for n not too large, the first source of loss decreases as n increases because both $N - n$ and $q(N - n, n)$ decrease. At the same time the second source of loss increases. By making a tradeoff between the first and second sources of loss, then, it is possible to find for each N a value $n^*(N)$ for which the losses are minimized; i.e.,

$$L(N - n^*, n^*) \leqq L(N - n, n) \quad \text{for all } n \leqq N.$$

For the determination of n^* let us assume that *initially* one random variable is as likely as the other to be best. (This would be the case for example if the two unbalanced coins referred to earlier have no identifying external characteristics and are positioned initially at random. More generally, the result is the same if the labels of the random variables are assigned at random. The proof of the theorem will indicate the modifications necessary for cases where one random variable is initially more likely than the other to be the best.) For convenience let us adopt the convention that ξ_1 is the random variable with the highest mean and let μ_1 be that mean; accordingly ξ_2 is the other random variable with mean $\mu_2 \leqq \mu_1$. (The observer, of course, does not know this.) Using these conventions we can now establish

THEOREM 5.1: *Given N trials to be allocated to two random variables, with means* $\mu_1 > \mu_2$ *and variances* σ_1^2, σ_2^2 *respectively, the minimum expected loss results when the number of trials allocated* $\xi_{(2)}(N)$ *is*

$$n \leqq n^* \sim b^2 \ln [N^2/(8\pi b^4 \ln N^2)]$$

where $b = \sigma_1/(\mu_1 - \mu_2)$. If, initially, one random variable is as likely as the other to be best, $n = n^*$ and the expected loss per trial is

$$L^*(N) \sim (b^2(\mu_1 - \mu_2)/N)[2 + \ln [N^2/(8\pi b^4 \ln N^2)]].$$

(Given two arbitrary functions, $Y(t)$ and $Z(t)$, of the same variable t, "$Y(t) \sim Z(t)$" will be used to mean $\lim_{t\to\infty} (Y(t)/Z(t)) = 1$ while "$Y(t) \cong Z(t)$" means that under stated conditions the difference $(Y(t) - Z(t))$ is negligible.)

Proof: In order to select an n which minimizes the expected loss, it is necessary first to write $q(N - n, n)$ as an explicit function of n. As defined above $q(N - n, n)$ is the probability that $\xi_{(2)}(N) = \xi_1$. More carefully, given the observation, say, that $\xi' = \xi_{(2)}(N)$, we wish to determine the probability that $\xi' = \xi_1$. That is, we wish to determine

$$q(N - n, n) = Pr\{\xi' = \xi_1 \mid \xi_{(2)} = \xi'\}$$

as an explicit function of $N - n$ and n. Bayes's theorem then gives us the equation

$$Pr\{\xi' = \xi_1 \mid \xi_{(2)} = \xi'\}$$
$$= \frac{Pr\{\xi' = \xi_{(2)} \mid \xi' = \xi_1\}Pr\{\xi' = \xi_1\}}{Pr\{\xi' = \xi_{(2)} \mid \xi' = \xi_1\}Pr\{\xi' = \xi_1\} + Pr\{\xi' = \xi_{(2)} \mid \xi' = \xi_2\}Pr\{\xi' = \xi_2\}}$$

Letting q', q'', and p designate $Pr\{\xi' = \xi_{(2)} \mid \xi' = \xi_1\}$, $Pr\{\xi' = \xi_{(2)} \mid \xi' = \xi_2\}$, and $Pr\{\xi' = \xi_1\}$, respectively, and using the fact that ξ' must be ξ_2 if it is not ξ_1, this can be rewritten as

$$q(N - n, n) = q'p/(q'p + q''(1 - p)).$$

(If one random variable is as likely as the other to be best, then $p = (1 - p) = \frac{1}{2}$.)

To derive q' let us assume that ξ' has received n trials out of the N total. Let S_2^{N-n} be the sum of the outcomes (payoffs) of $N - n$ trials of ξ_2 and let S_1^n be the corresponding sum for n trials of ξ_1. Since q' has $\xi' = \xi_1$ as a condition, q' is just the probability that $S_1^n/n < S_2^{N-n}/(N - n)$ or, equivalently the probability that $(S_1^n/n) - (S_2^{N-n}/(N - n)) < 0$. By the central limit theorem $S_2^{N-n}/(N - n)$ approaches a normal distribution with mean μ_2 and variance $\sigma_2^2/(N - n)$; similarly, S_1^n/n has mean μ_1 and variance σ_1^2/n. The distribution of $(S_1^n/n) - (S_2^{N-n}/(N - n))$ is given by the product (convolution) of the distributions of S_1^n/n and $-(S_2^{N-n}/(N - n))$; by an elementary theorem (on the convolution of normal distributions) this is a normal distribution with mean $\mu_1 - \mu_2$ and variance $\frac{\sigma_1^2}{n} + \frac{\sigma_2^2}{N - n}$. Thus the probability $Pr\left\{\frac{S_1^n}{n} - \frac{S_2^{N-n}}{N - n} < 0\right\}$ is the tail $1 - \Phi(x_0)$ of a canonical normal distribution $\Phi(x)$ where

$$x = \frac{y - (\mu_1 - \mu_2)}{\sqrt{\dfrac{\sigma_1^2}{n} + \dfrac{\sigma_2^2}{N - n}}}$$

and $-x_0$ is the value of x when $y = 0$. (I.e., $\Phi(y)$, which describes the distribution of $\dfrac{S_1^n}{n} - \dfrac{S_2^{N-n}}{N-n}$, is transformed to $\Phi(x)$ which describes the canonical normal distribution with mean 0 and variance 1.) The tail of a normal distribution is well approximated by

$$\Phi(-x) - 1 - \Phi(x) \lesssim \frac{1}{\sqrt{2\pi}} \cdot \frac{e^{-x^2/2}}{x}.$$

Thus

$$q' \lesssim \frac{1}{\sqrt{2\pi}} \cdot \frac{e^{-x_0^2/2}}{x_0} = \frac{1}{\sqrt{2\pi}} \frac{\sqrt{\dfrac{\sigma_1^2}{n} + \dfrac{\sigma_2^2}{N - n}}}{(\mu_1 - \mu_2)} \exp \frac{1}{2} \left[\frac{-(\mu_1 - \mu_2)^2}{\dfrac{\sigma_1^2}{n} + \dfrac{\sigma_2^2}{N - n}} \right].$$

Using the same line of reasoning (but now with $(N - n)$ observations of ξ_1, etc.) we have

$$q'' = 1 - Pr\left\{ \frac{S_1^{N-n}}{N - n} < \frac{S_2^n}{n} \right\}$$

$$\lesssim 1 - \frac{1}{\sqrt{2\pi}} \frac{\sqrt{\dfrac{\sigma_1^2}{N - n} + \dfrac{\sigma_2^2}{n}}}{(\mu_1 - \mu_2)} \exp \frac{1}{2} \left[\frac{-(\mu_1 - \mu_2)^2}{\dfrac{\sigma_1^2}{N - n} + \dfrac{\sigma_2^2}{n}} \right].$$

From this we see that both q' and q'' are functions of the variances and means as well as the total number of trials, N, and the number of trials, n, given ξ'. More importantly, both q' and $1 - q''$ decrease exponentially with n, yielding

$$q(N - n, n) = q'p/(q'p + q''(1 - p)) \sim q' \cdot (p/(1 - p))$$

with the approximation being quite good even for relatively small n. For $p = \frac{1}{2}$ this reduces to

$$q(N - n, n) \sim q'$$

where the error is less than min $\{(q')^2, (1 - q'')^2\}$. (If one random variable is a priori more likely than the other to be best, i.e., if $p \neq \frac{1}{2}$, then we can see from

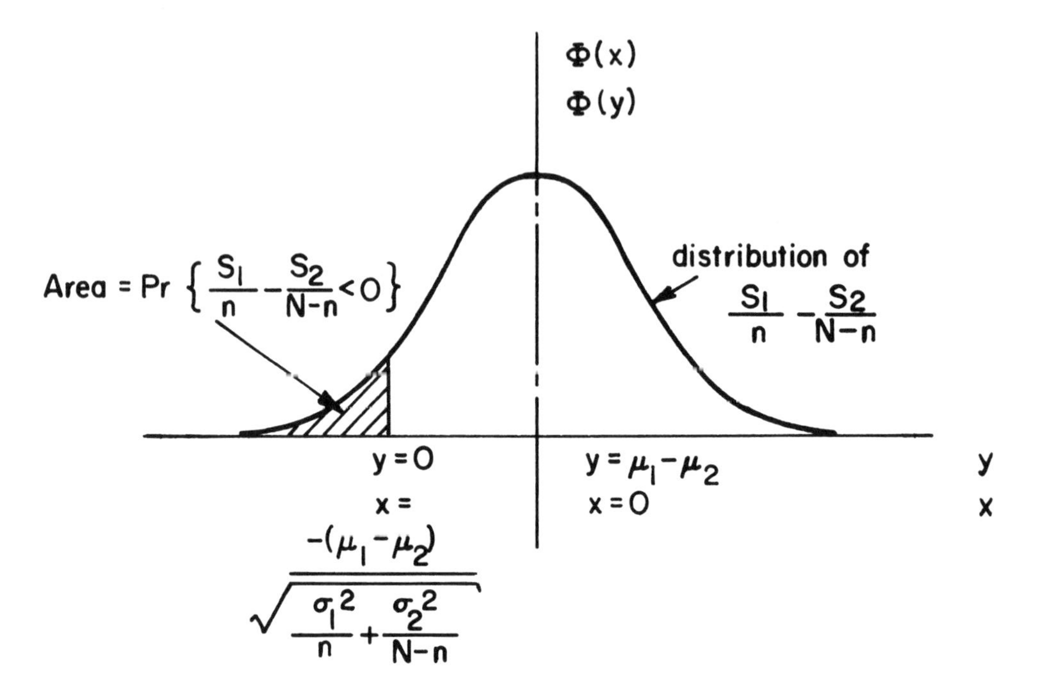

Fig. 11. The convolution of $\frac{S_1}{n}$ *with* $\frac{S_2}{N-n}$

above and from what follows that fewer trials can be allocated to attain the same reduction of $q(N - n, n)$. The expected loss is reduced accordingly.)

The observation that q' and hence $q(N - n, n)$ decreases exponentially with n makes it clear that, to minimize loss as N increases, the number of trials allocated the observed best, $N - n$, should be increased dramatically relative to n. This observation (which will be verified in detail shortly) enables us to simplify the expression for x_0. Whatever the value of σ_2, there will be an N_0 such that, for any $N > N_0$, $\sigma_2^2/(N - n) \ll \sigma_1^2/n$, for n close to its optimal value. (In most cases of interest this occurs even for small numbers of trials since, usually, σ_1 is at worst an order of magnitude or two larger than σ_2.) Using this we see that, for n close to its optimal value,

$$x_0 \lesssim \frac{(\mu_1 - \mu_2)\sqrt{n}}{\sigma_1}, \; N > N_0.$$

We can now proceed to determine what value of n will minimize the loss $L(n)$ by taking the derivative of L with respect to n.

$$\frac{dL}{dn} = |\mu_1 - \mu_2| \cdot \left[-q + (N-n)\frac{dq}{dn} + 1 - q - n\frac{dq}{dn}\right]$$
$$= |\mu_1 - \mu_2| \cdot \left[(1-2q) + (N-2n)\frac{dq}{dn}\right]$$

where

$$\frac{dq}{dn} \lesssim \frac{1}{\sqrt{2\pi}}\left[-\frac{e^{-x_0^2/2}}{x_0^2} - e^{-x_0^2/2}\right]\frac{dx_0}{dn} = -\left[\frac{q}{x_0} + x_0 q\right]\frac{dx_0}{dn}$$

and

$\dfrac{dx_0}{dn} \lesssim \dfrac{\mu_1 - \mu_2}{2\sigma_1\sqrt{n}} = \dfrac{x_0}{2n}$. Thus $\dfrac{dL}{dn} \lesssim |\mu_1 - \mu_2| \cdot \left[(1-2q) - (N-2n)\cdot q \cdot \dfrac{(x_0^2+1)}{2n}\right]$.

n^*, the optimal value of n, satisfies $\dfrac{dL}{dn} = 0$, whence we obtain a bound on n^* as follows:

$$0 \lesssim (1-2q) - \left(\frac{N-2n^*}{2n^*}\right) \cdot q\cdot(x_0^2+1)$$

or

$$\left(\frac{N-2n^*}{2n^*}\right) \lesssim \frac{1-2q}{q\cdot(x_0^2+1)}$$

Noting that $1/(x_0^2+1) \lesssim 1/x_0^2$ and that $(1-2q)$ rapidly approaches 1 because q decreases exponentially with n, we see that $\dfrac{N-2n^*}{n^*} \lesssim \dfrac{2}{x_0^2 q}$, where the error rapidly approaches zero as N increases. Thus the observation of the preceding paragraph is verified, the ratio of trials of the observed best to trials of second-best growing exponentially.

Finally, to obtain n^* as an explicit function of N, q must be written in terms of n^*:

$$\frac{N-2n^*}{n^*} \lesssim \frac{2\sqrt{2\pi}\sigma_1}{(\mu_1-\mu_2)} \cdot \frac{1}{\sqrt{n^*}} \cdot \exp\left[((\mu_1-\mu_2)^2 n^*)/(2\sigma_1^2)\right].$$

Introducing $b = \sigma_1/(\mu_1 - \mu_2)$ and $N_1 = N - n^*$ for simplification, we obtain

$$N_1 \lesssim \sqrt{8\pi}\cdot b\cdot \exp\left[(b^{-2}n^* + \ln n^*)/2\right]$$

or

$$n^* + b^2 \ln n^* \gtrsim 2b^2 \cdot \ln \frac{N_1}{b\sqrt{8\pi}}$$

where the fact that $(N - 2n^*) \sim (N - n^*)$ has been used, with the inequality generally holding as soon as N_1 exceeds n^* by a small integer. We obtain a recursion for an ever better approximation to n^* as a function of N_1 by rewriting this as

$$n^* \gtrsim b^2 \ln \left[\frac{(b^{-1}N_1)^2}{8\pi n^*}\right].$$

Whence

$$\begin{aligned} n^* &\gtrsim b^2 \ln \left[\frac{(b^{-1}N_1)^2}{8\pi(b^2 \ln ((b^{-1}N_1)^2/8\pi n^*))}\right] \\ &\gtrsim b^2 \ln \left[\frac{b^{-4}N_1^2}{8\pi} \cdot \frac{1}{\ln ((b^{-1}N_1)^2/8\pi) - \ln n^*}\right] \\ &\gtrsim b^2 \ln \left[\frac{b^{-4}N_1^2}{8\pi(\ln N_1^2 - \ln (b^{-2}/8\pi))}\right] \\ &\gtrsim b^2 \ln \left[\frac{b^{-4}N_1^2}{8\pi \ln N_1^2}\right], \end{aligned}$$

where, again, the error rapidly approaches zero as N increases. Finally, where it is desirable to have n^* approximated by an explicit function of N, the steps here can be redone in terms of N/n^*, noting that N_1/n^* rapidly approaches N/n^* as N increases. Then

$$n^* \sim b^2 \ln \left[\frac{N^2}{8\pi b^4 \ln N^2}\right]$$

where, still, the error rapidly approaches zero as N increases.

The expected loss per trial $L^*(N)$ when n^* trials have been allocated to $\xi_{(2)}(\tau, N)$ is

$$\begin{aligned} L^*(N) &= \frac{1}{N}|\mu_1 - \mu_2| \cdot [(N - n^*)q(N - n, n^*) + n^*(1 - q(N - n^*, n^*))] \\ &= |\mu_1 - \mu_2| \cdot \left[\frac{N - 2n^*}{N} q(N - n^*, n^*) + \frac{n^*}{N}\right] \\ &\gtrsim |\mu_1 - \mu_2| \cdot \left[\frac{2n^*}{Nx_0^2} + \frac{n^*}{N}\right] \\ &\gtrsim \frac{b^2|\mu_1 - \mu_2|}{N} \cdot \left[2 + \ln\left(\frac{N^2}{8\pi b^4 \ln N^2}\right)\right]. \end{aligned}$$

Q.E.D.

The main point of this theorem quickly becomes apparent if we rearrange the results to give the number of trials $N^*_{(1)}$ allocated to $\xi_{(1)}$ as a function of the number of trials n^* allocated to $\xi_{(2)}$:

$$N^*_{(1)} = N - n^* \sim N \sim \sqrt{8\pi b^4 \ln N^2}\, e^{n^*/2b^2}.$$

Thus the loss rate will be optimally reduced if the number of trials allocated $\xi_{(1)}$ grows slightly faster than an exponential function of the number of trials allocated $\xi_{(2)}$. *This is true regardless of the form of the distributions defining ξ_1 and ξ_2.* Later we will see that the random variables defined by schemata are similarly treated by reproductive plans.

It should be emphasized that the above approximation for n^* will be misleading for small N when

(i) $\mu_1 - \mu_2$ is small enough that, for small N, the standard deviation of $\frac{S_1}{n} - \frac{S_2}{N-n}$ is large relative to $\mu_1 - \mu_2$ and, as a consequence, the approximation for the tail $1 - \Phi(x_0)$ fails,

or (ii) σ_2 is large relative to σ_1 so that, for small N, the approximation for x_0 is inadequate.

Neither of these cases is important for our objectives here. The first is unimportant because the cumulative losses will be small until N is large since the cost of trying ξ_2 is just $\mu_1 - \mu_2$. The second is unimportant because the uncertainty and therefore the expected loss depends primarily on ξ_1 until $N - n^*$ is large; hence the expected loss rate will be reduced near optimally as long as $N - n \cong N$ (i.e., most trials go to $\xi_{(1)}$), as will be the case if n is at least as small as the value given by the approximation for n^*.

Finally, to get some idea of n^* when σ_1 is not known, note that for bounded payoff, $\xi_i:\mathcal{A} \to [r_0, r_1]$, the maximum variance occurs when all payoff is concentrated at the extremes, i.e., $p(r_0) = p(r_1) = \frac{1}{2}$. Then

$$\sigma_1^2 \leqq \sigma_{\max}^2 = \left(\frac{1}{2}r_1^2 + \frac{1}{2}r_0^2\right) - \left(\frac{1}{2}r_1 + \frac{1}{2}r_0\right)^2 = \left(\frac{r_1 - r_0}{2}\right)^2.$$

2. REALIZATION OF MINIMAL LOSSES

This section points out, and resolves, a difficulty in using $L^*(N)$ as a performance criterion. The difficulty occurs because, in a strict sense, the minimal expected loss rate just calculated cannot be obtained by any feasible plan for allocating trials in terms of observations. As such $L^*(N)$ constitutes an unattainable lower bound and, if it is too far below what *can* be attained, it will not be a useful criterion. However, we will see here that such loss rates can be approached quite closely (arbitrarily closely as N increases) by feasible plans, thus verifying $L^*(N)$'s usefulness.

The source of the difficulty lies in the expression for n^*, which was obtained on the assumption that the n^* trials were allocated to $\xi_{(2)}(N)$. However there is *no* realizable plan (sequential algorithm) which can "foresee" in all cases which of the two random variables will be $\xi_{(2)}(N)$ at the end of N trials. No matter what the plan τ, there will be some observational sequences for which it allocates $n > n^*$ trials to a random variable ξ (on the assumption that ξ will be $\xi_{(1)}(N)$) only to have ξ turn out to be $\xi_{(2)}(N)$ after N trials. (For example, the observational sequence may be such that at the end of $2n^*$ trials τ has allocated n^* trials to each random variable. τ must then decide where to allocate the next trial even though each random variable has a positive probability of being $\xi_{(2)}(N)$.) For these sequences the loss rate will perforce exceed the optimum. Hence $L^*(N)$ is not attainable by any realizable τ—there will always be payoff sequences which lead τ to allocate too many trials to $\xi_{(2)}(N)$.

There is, however, a realizable plan $\tau_{(\sim)}$ for which the expected loss per trial $L(\tau_{(\sim)}, N)$ quickly approaches $L^*(N)$, i.e.,

$$\lim_{N\to\infty} (L(\tau_{(\sim)}, N)/L^*(N)) = 1.$$

$\tau_{(\sim)}$ proceeds by initially allocating n^* trials to each random variable (in any order) and then allocates the remaining $N - 2n^*$ trials to the random variable with the highest observed payoff rate at the end of the $2n^*$ trials. It is important to note that n^* is *not* recalculated for $\tau_{(\sim)}$; it is the value determined above.

COROLLARY 5.2: *Given N trials, $\tau_{(\sim)}$'s expected loss $L(\tau_{(\sim)}, N)$ approaches the optimum $L^*(N)$; i.e., $L(\tau_{(\sim)}, N) \sim L^*(N)$.*

Proof: The expected loss per trial $L(\tau_{(\sim)}, N)$ for $\tau_{(\sim)}$ is determined by applying the earlier discussion of sources of loss to the present case.

$$L(\tau_{(\sim)}, N) = \frac{1}{N} \cdot (\mu_1 - \mu_2) \cdot [(N - n^*)q(n^*, n^*) + n^*(1 - q(n^*, n^*))]$$

where q is the same function as before, but here the probability of error is irrevocably determined after $2n^*$ trials. That is

$$q(n^*, n^*) \sim \left(\sqrt{\frac{\sigma_1^2}{n^*} + \frac{\sigma_2^2}{n^*}} \Big/ \sqrt{2\pi}(\mu_1 - \mu_2)\right) \cdot \exp \frac{1}{2}\left[-(\mu_1 - \mu_2)^2 \Big/ \left(\frac{\sigma_1^2}{n^*} + \frac{\sigma_2^2}{n^*}\right)\right].$$

Rewriting $L(\tau_{(\sim)}, N)$ we have

$$L(\tau_{(\sim)}, N) = (\mu_1 - \mu_2)\left[\frac{N - 2n^*}{N}\, q(n^*, n^*) + \frac{n^*}{N}\right].$$

Since, asymptotically, q decreases as rapidly as N^{-1}, it is clear that the second term in the brackets will dominate as N grows. Inspecting the earlier expression for $L(N)$ we see the same holds there. Thus, since the second terms are identical

$$\lim_{N\to\infty} (L(\tau_{(\sim)}, N)/L^*(N)) = 1. \qquad \text{Q.E.D.}$$

From this we see that, given the requisite information (μ_1, σ_1) and (μ_2, σ_2), there exist plans which have loss rates closely approximating $L^*(N)$ as N increases.

3. MANY OPTIONS

The function $L^*(N)$ sets a very stringent criterion when there are *two* uncertain options, specifying a high goal which can only be approached where uncertainty is very limited. Adaptive plans, however, considered in terms of testing schemata, face many more than two uncertain options at any given time. Thus a general performance criterion for adaptive plans must treat loss rates for arbitrary numbers of options. Though the extension from two options to an arbitrary number of r options is conceptually straightforward, the actual derivation of $L^*(N)$ is considerably more intricate. The derivation proceeds by indexing the r random variables $\xi_1, \xi_2, \ldots, \xi_r$ so that the means are in decreasing order $\mu_1 > \mu_2 > \cdots > \mu_r$ (again, without the observer knowing that this ordering holds).

THEOREM 5.3: *Under the same conditions as for Theorem 5.1, but now with r random variables, the minimum expected loss after N trials must exceed*

$$(\mu_1 - \mu_2)\cdot(r - 1)b^2\left[2 + \ln\left(\frac{N^2}{8\pi(r-1)^2 b^4 \ln N^2}\right)\right]$$

where $b = \sigma_1/(\mu_1 - \mu_r)$.

Proof: Following Theorem 5.1 we are interested in the probability that the average of the observations of any ξ_i, $i > 1$, exceeds the average for ξ_1. This probability of error is accordingly

$$q(n_1, \ldots, n_r) \sim Pr\left\{\left(\frac{S_1}{n_2} < \frac{S_2}{n_1}\right) \text{ or } \left(\frac{S_1}{n_3} < \frac{S_3}{n_1}\right) \text{ or} \ldots \text{or } \left(\frac{S_1}{n_r} < \frac{S_r}{n_1}\right)\right\},$$

where n_i is the number of trials given ξ_i, and the loss ranges from $(\mu_1 - \mu_2)$ to $(\mu_1 - \mu_r)$ depending on which ξ_i is mistakenly taken for best.

Let $n = \sum_{i=2}^{r} n_i$, let $m = \min \{n_2, n_3, \ldots, n_r\}$, and let j be the largest index of the random variables (if more than one) receiving m trials.

The proof of Theorem 5.1 shows that a lower bound on the expected loss is attained by minimizing with respect to any lower bound on the probability q (a point which will be verified in detail for r variables). In the present case q must exceed

$$Pr\left\{\frac{S_1}{m} < \frac{S_j}{N-n}\right\} > q' = \frac{1}{\sqrt{2\pi}} \frac{\sigma_1}{(\mu_1 - \mu_r)\sqrt{m}} \exp\left[\frac{(\mu_1 - \mu_r)^2 m}{2\sigma_1^2}\right]$$

using the fact that $(\mu_1 - \mu_r) \geq (\mu_1 - \mu_j)$ for any $j > 1$. By the definition of q

$$L_{N,r}(n) > L'_{N,r}(n) = (\mu_1 - \mu_2)[(N-n)q + n(1-q)]$$

using the fact that $(\mu_1 - \mu_2) \leqq (\mu_1 - \mu_i)$ for $i \geqq 2$. Moreover the same value of n minimizes both $L_{N,r}(n)$ and $L'_{N,r}(n)$. To find this value of n, set

$$\frac{dL'_{N,r}}{dn} = (\mu_1 - \mu_2)\left[(N - 2n)\frac{dq}{dn} + (1 - 2q)\right] = 0.$$

Solving this for n^* and noting that $1 - 2q$ rapidly approaches 1 as N increases, gives

$$n^* \sim \frac{N}{2} + \left(\frac{dq}{dn}\right)^{-1}.$$

Noting that q must decrease less rapidly than q' with increasing n, we have $(dq'/dn) < (dq/dn)$ and, taking into account the negative sign of the derivatives,

$$n^* > \frac{N}{2} + \left(\frac{dq'}{dn}\right)^{-1}.$$

(This verifies the observation at the outset, since the expected loss approaches n^* as N increases—see below.) Finally, noting that $n > (r-1)m$, we can proceed as in the two-variable case by using $(r-1)m$ in place of n and taking the derivative of q' with respect to m instead of n. The result is

$$n^* > (r-1)m^* \sim (r-1)b^2 \ln\left(\frac{N^2}{8\pi(r-1)^2 b^4 \ln N^2}\right)$$

where $b = \sigma_1/(\mu_1 - \mu_r)$. Accordingly,

$$L_{N,r}(n^*) > L'_{N,r}((r-1)m^*)$$
$$> (\mu_1 - \mu_2)\cdot(r-1)b^2\left[2 + \ln\left(\frac{N^2}{8\pi(r-1)^2b^4 \ln N^2}\right)\right]. \quad \text{Q.E.D.}$$

4. APPLICATION TO SCHEMATA

We are ready now to apply the criterion just developed to the general problem of ranking schemata. The basic problem was rephrased as one of minimizing the performance losses inevitably coupled with any attempt to increase confidence in an observed ranking of schemata. The theorem just proved provides a guideline for solving this problem by indicating how trials should be allocated among the schemata of interest.

To see this note first that the central limit theorem, used at the heart of the proof of Theorem 5.3, applies to any sequence of independent random variables having means and variances. As such it applies to the observed average payoff $\hat{\mu}_\xi$ of a sequence of trials of the schema ξ under any probability distribution P over $\mathcal{A}$ (cf. chapter 4). It even applies when the distribution over $\mathcal{A}$ changes with time (a fact we will take advantage of with reproductive plans). In particular, then, Theorem 5.3 applies to any given set of r schemata. It indicates that under a good adaptive plan the number of trials of the (observed) best will increase exponentially relative to the total number of trials allocated to the remainder.

Near the end of chapter 4 it was proposed that the observed performance rankings of schemata be stored by selecting an appropriate (small) set of elements $\mathcal{B}$ from $\mathcal{A}$ so that the rank of each schema would be indicated by the relative number of instances of ξ in $\mathcal{B}$. Theorem 5.3 suggests an approach to developing $\mathcal{B}$, or rather a sequence $\mathcal{B}(1), \mathcal{B}(2), \ldots, \mathcal{B}(t)$, according to the sequence of observations of schemata. Let the number of instances of ξ in the set $\mathcal{B}(t)$ represent the number of observations of ξ at time t. Then the number of instances of ξ in the set $\bigcup_{t=1}^{T} \mathcal{B}(t)$ represents the total number of observations of ξ through time T. If schema ξ should persist as the observed best, Theorem 5.3 indicates that ξ's portion of $\bigcup_{t=1}^{T} \mathcal{B}(t)$ should increase exponentially with respect to the remainder. We can look at this in a more "instantaneous" sense. ξ's portion of $\mathcal{B}(t)$ corresponds to the rate at which ξ is being observed, i.e., to the "derivative" of the function giving ξ's increase. Since the derivative of an exponential is an exponential, it seems natural to have ξ's portion $M_\xi(t)$ of $\mathcal{B}(t)$ increase exponentially with t (at least until ξ

occupies most of $\mathcal{B}(t)$). This will be the case if ξ's rate of increase is proportional to the observed average payoff $\hat{\mu}_\xi(t)$ of instances of ξ at time t or, roughly,

$$dM_\xi(t)/dt = \hat{\mu}_\xi(t)M_\xi(t).$$

It will still be the case if the rate is proportional to the schema's "usefulness," the difference between $\hat{\mu}_\xi(t)$ and the overall average performance $\hat{\mu}(t)$ of instances in $\mathcal{B}(t)$, so that $dM_\xi(t)/dt = (\hat{\mu}_\xi(t) - \hat{\mu}(t))M_\xi(t)$. (In genetics $\hat{\mu}_\xi(t) - \hat{\mu}(t)$ is called the "average excess" of ξ when ξ is defined on a single locus, i.e., when ξ is a specific allele.)

The discussion of "intrinsic parallelism" in chapter 4 would imply here that *each* ξ represented in $\mathcal{B}(t)$ should increase (or decrease) at a rate proportional to *its* observed "usefulness" $\hat{\mu}_\xi(t) - \hat{\mu}(t)$. If this could be done consistently then each ξ would be automatically and properly ranked within $\mathcal{B}(t)$ as t increases. The reasoning behind this, as well as the proof that reproductive plans accomplish the task, will be developed in full in the next two chapters.

6. Reproductive Plans and Genetic Operators

In the earlier informal discussion of genetics (sections 1.4 and 3.1) reproductive plans were introduced as the fundamental procedure of genetic adaptation. The present chapter lifts reproductive plans from the specific context of genetics to the general framework of chapter 2. This, at one stroke, makes reproductive plans suitable objects for rigorous study *and* yields a class of plans applicable to the full range of adaptive systems. Genetic plans, i.e., reproductive plans using generalized genetic operators, will be the prime focus; emphasis will be laid upon the operators' retention and use of relevant history as they exploit opportunities for improved performance.

Genetic plans can be applied to any domain of structures $\mathcal{A}$ represented by strings (l-tuples). (To build a better intuition for this flexibility the reader may find it useful to consistently interpret the properties and theorems advanced here in the most familiar of the nongenetic illustrations of chapter 3.) We will see that *each* structure generated and tested by a genetic plan in effect tests a multitude of schemata and that the plan actually preserves and exploits this information. Genetic plans do this by generating sequences of structures in such a way that, once a few instances of *any* given schema ξ occur, one can count on the cumulative number of instances of ξ increasing at a rate closely related to μ_ξ. The generalized genetic operators act so as to test old schemata in new contexts, generate instances of schemata not previously tested, and so on (see sections 7.2–7.5), without disturbing the rates of increase. Genetic plans thus exhibit the intrinsic parallelism discussed at the ends of chapters 4 and 5.

Interpreted in genetics, the results of the next two chapters indicate that adaptation proceeds largely in terms of pools of coadapted sets of alleles rather than gene pools. As one important offshoot, this approach yields an extension of Fisher's (1930) classical result (on the rates of increase of alleles) to coadapted sets of alleles with epistatic interaction (see section 7.4). A typical interpretation for artificial systems can be obtained by looking again at the function $f(x)$ of Figure 10.

We see that the average value $\mu_{1\square\square\ldots\square}$ of all points in the schema $1\square\square\ldots\square$ (i.e., the area under the curve over the interval $\frac{1}{2} \leqq x < 1$ divided by $\frac{1}{2}$, the length of the interval) is approximately 1.5. Similarly, for $\square\square0\square\ldots\square$ the value is approximately 1, for $1\square0\square\ldots\square$ the value is approximately 2, etc. Thus instances of $1\square\square\ldots\square$ will accumulate at a higher rate than those of $\square\square0\square\ldots\square$, and instances of $1\square0\square\ldots\square$ will accumulate still more rapidly. The result is an ever greater clustering of test points (instances) in intervals (schemata) of above-average value (see Figure 13 and the example of section 7.3). In this way the genetic plan locates a global optimum of $f(x)$, exploiting false peaks (without entrapment) to rapidly increase the average value of points tested.

We will see that genetic plans act with a combination of simplicity and subtlety both pleasing to the eye and useful in application. They also act with robustness and efficiency, a fact that will be finally established in the next chapter. It should be emphasized that the plans (algorithms) set forth have a dual role. When the plan's parameter values (and functions) are determined from data about a particular natural process, the plan serves as an idealized model or hypothesis about that process. As such it is subject to the general observation-modification cycle applicable to physical theories in general. Because the model is already in algorithmic form, it is particularly suitable for simulations of the process. The other role occurs in relation to artificial (designed) processes. Here the plans serve as optimization procedures which can be fitted into the process to control its direction. In either role the theorems proved hereafter yield predictions which must come true if (for the natural systems) the basic model is verified or (for artificial systems) the algorithm is incorporated as a control.

1. GENERALIZED REPRODUCTIVE PLANS

To embed reproductive plans in the $(\mathfrak{I}, \mathcal{E}, \chi)$ framework of chapter 2 we must define a class of plans (algorithms) applicable to an arbitrary set of structures $\mathcal{A}$. Moreover each plan must be a mapping of the form $\tau: I \times \mathcal{A} \rightarrow \Omega$. It must use only the input from the environment, $I(t)$, and the structure tried at time t, $\mathcal{A}(t)$, to determine a random variable over $\mathcal{A}$, $\omega_t(\mathcal{A}(t))$, which is in turn sampled to determine the next trial, $\mathcal{A}(t+1)$. We will begin by defining a relatively narrow class of reproductive plans $\mathcal{R}_1$. Later $\mathcal{R}_1$ will be extended in ways which make some applications more natural, and we will see that the new algorithms are essentially no more powerful than those from $\mathcal{R}_1$.

To begin let $\mathcal{A}_1$ be the set of structures to be tested and, as in chapter 4, assume that the elements of $\mathcal{A}_1$ are representations. (As long as each structure is represented by a finite string of attributes, $\mathcal{A}_1$ can be made countably infinite without affecting the presentation of $\mathcal{R}_1$. This will be discussed in chapter 8.) Each plan in $\mathcal{R}_1$ is an algorithm which acts at each instant t upon a small *set* of structures $\mathcal{B}(t)$ from $\mathcal{A}_1$ (interpretable, for instance, as a population or data base). The algorithm uses a single basic cycle to modify elements of the small set, one at a time, thereby producing a sequence of new structures for trial. In general terms, the basic steps of the cycle are:

1. Select one structure from $\mathcal{B}(t)$ probabilistically, after assigning each structure a probability proportional to its observed performance.
2. Copy the selected structure, then apply operators to the copy to produce a new structure.
3. Select a second element from $\mathcal{B}(t)$ at random (all elements equally likely) and replace it by the new structure produced in step 2.
4. Observe and record the performance of the new structure.
5. Return to step 1.

Note that the *number* of elements in $\mathcal{B}(t)$ remains constant. (From the point of view of genetics, it is convenient to look upon the size of $\mathcal{B}(t)$ as an upper bound on population size determined, say, by the "carrying capacity" of the environment.) The number of structures in $\mathcal{B}(t)$ can be varied up to the maximum number by allowing null structures or vacancies.

With this outline as a guide, we can now go on to the rigorous definition of the algorithms in $\mathcal{R}_1$. The following symbols and definitions will be used with the interpretations given:

$\mathcal{A}_1$,	the set of basic structures being tested.
$\mathcal{A}_1^M$,	the set of all M-tuples of structures corresponding to possible compositions of $\mathcal{B}$.
$\mathcal{B}(t)$,	the particular set of M structures $\{A_1(t), A_2(t), \ldots, A_M(t)\}$ available to the adaptive plan at time t.
$\mathcal{I}_M = \{1, 2, \ldots, M\}$,	the first M positive integers, used as an index set for $\mathcal{B}$.
Ω,	the set of stochastic operators for modifying structures.
$\mathcal{I}_M \times \mathcal{A}_1^M$,	compositions of $\mathcal{B}$ with one structure selected (for modification by an operator); i.e. $(i, A_1(t), \ldots, A_M(t)) \in \mathcal{I}_M \times \mathcal{A}_1^M$ corresponds to $\mathcal{B}(t)$ with the ith structure, $A_i(t)$ selected.

$\mathcal{P}$,	a set of probability distributions over $\mathcal{A}_1$, one of which is selected by each application of a stochastic operator $\omega \in \Omega$.
$\rho: \mathcal{A}_1 \to \Omega$,	assigns to each basic structure $A \in \mathcal{A}_1$ the stochastic operator $\omega \in \Omega$ which is to be used to modify A.
$\omega: \mathcal{I}_M \times \mathcal{A}_1^M \to \mathcal{P}$,	an arbitrary operator from Ω which determines, from $\mathcal{B}(t)$ and a selection $i(t)$, a distribution $P \in \mathcal{P}$ over $\mathcal{A}_1$.

Once the set of structures $\mathcal{A}_1$ has been given, along with an observation procedure which assigns a payoff $\mu_E(A)$ to each trial of a structure $A \in \mathcal{A}_1$, a reproductive plan of type $\mathcal{R}_1$ is determined by specifying the functions ρ and $\{\omega\}$. The algorithm proceeds as follows:

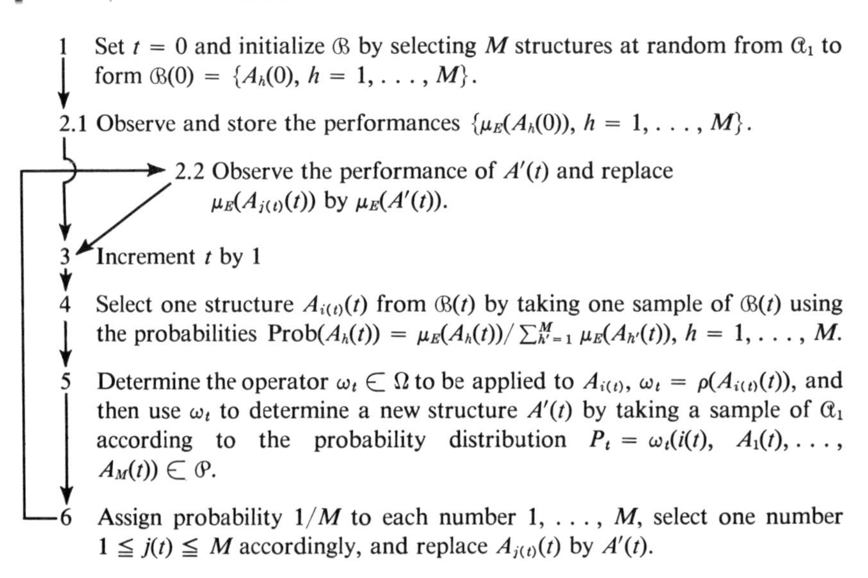

Algorithms of type $\mathcal{R}_1$ are strictly sequential in the sense that one individual $A'(t)$ is tested at a time. $\mathcal{B}(t)$ serves as a reservoir of information about the environment and as a basis for generating new trials. $\mathcal{B}(t)$ remains constant in size because each new individual $A'(t)$ replaces an individual already in the population. Under the operators Ω of interest (particularly the generalized genetic operators), $A'(t)$ can be looked upon as the "offspring" of $A_{i(t)}(t)$, retaining many (but generally not all) of the attributes of $A_{i(t)}(t)$. Via the function ρ each structure in the population carries a specification of the operator appropriate to it (a kind of "species" designation). (The apparent generalization to stochastic selection of one of a *set* of operators can actually be subsumed in the stochastic selection of offspring. See below.) The operators are computation procedures using random numbers; generally, they use at most one other member of the population, in addition to $A_{i(t)}(t)$, in the determination of $A'(t)$. (For instance, the operator may randomly select a "mate" for $A_{i(t)}(t)$.) The argument of each $\omega \in \Omega$ includes the whole population, because any structure in the population is a conceivable candidate for the second operand, even when ω is essentially a binary operator. (E.g., the probable outcomes of a "mating" will depend upon the range of "mates" available.)

It should be noted that the *state* of the algorithm at the beginning of any cycle includes not only the population $\mathcal{B}(t)$, but also the retained performances $\mu_E(A_h(t))$, $h = 1, \ldots, M$, of the structures in $\mathcal{B}(t)$. Thus, in the general formalism of chapter 2,

$$\mathcal{A} = \mathcal{A}_1^M \times [0, r]^M$$

where $[0, r]$ is the interval of possible payoffs (performances), i.e. $[0, r]$ is the range of μ_E,

$$\mu_E : \mathcal{A}_1 \rightarrow [0, r].$$

Accordingly,

$$\mathcal{A}(t) = (A_1(t), \ldots, A_M(t), \mu_E(A_1(t)), \ldots, \mu_E(A_M(t))).$$

The *new* information $I(t)$, from the environment $E \in \mathcal{E}$ at each time t, is simply the payoff $\mu_E(A'(t))$ of the new structure $A'(t)$. Thus any adaptive plan $\tau \in \mathcal{R}_1$ has the required form

$$\tau : I \times \mathcal{A} \rightarrow \mathcal{A}$$

since

$$\begin{aligned} &\tau(\mu_E(A'(t)), [A_1(t), \ldots, A_M(t), \mu_E(A_1(t)), \ldots, \mu_E(A_M(t))]) \\ &\qquad = [A_1(t+1), \ldots, A_M(t+1), \mu_E(A_1(t+1)), \ldots, \mu_E(A_M(t+1))]. \end{aligned}$$

Informally, a reproductive plan is one under which the better an individual performs the more offspring it has. For plans $\tau \in \mathcal{R}_1$ a precise counterpart of this property can be established with the help of the following

LEMMA 6.1: *If, at any time-step, p_1 is the probability that a structure A produces an "offspring" during that time-step and p_2 is the probability that A is deleted during that time-step, then the expected number of "offspring" of A is p_1/p_2.*

Proof: This is immediately established by noting that, when p_1 and p_2 are constant, the expected lifespan of A is $1/p_2$ and the expected number of offspring is simply the number of offspring expected during the expected lifespan, i.e., p_1/p_2. In more detail, the probability of A surviving for *exactly* T time-steps is $p(T) = (1 - p_2)^{T-1} \cdot p_2$, and the expected number of "offspring" during that interval is $\bar{u}_A(T) = p_1 T$. Thus, the expected number of "offspring" during A's lifespan is

$$\sum_{T=1}^{\infty} p(T)\bar{u}_A(T) = p_1 p_2 \sum_{T=1}^{\infty} T(1 - p_2)^{T-1}.$$

But $\sum_{T=1}^{\infty} T(1 - p_2)^{T-1}$ converges to $(1/p_2)^2$ (as may be easily verified by taking the derivative of both sides of the identity $(1/1 - x) = 1 + x + x^2 + \ldots$). Therefore

$$\sum_{T=1}^{\infty} p(T)\bar{u}_A(T) = p_1/p_2. \qquad \text{Q.E.D.}$$

For plans in $\mathcal{R}_1$ the interpretation of this lemma is direct: The probability of A_h being selected to produce an offspring A' during time t is $\mu_{ht}/\sum_{h'}\mu_{h't}$ where $\mu_{h't} = {}^{df.}\mu_E(A_{h'}(t))$, while the probability of A_h being deleted at the end of that time-step is $1/M$. Hence, if $\sum_{h'}\mu_{h't}$ changes negligibly over A_h's lifespan, the expected number of offspring is

$$(\mu_{ht}/\sum_{h'}\mu_{h't})/(1/M) = \mu_{ht}/(\sum_{h'}\mu_{h't}/M) = \mu_{ht}/\bar{\mu}_t.$$

$\mu_{ht}/\bar{\mu}_t$ can be looked upon as a "normalized" payoff, the "usefulness" of A_h being measured relative to the average performance of the other members in the population. With this arrangement the expected number of offspring of A_h is greater than 1 just in case A_h's performance is above average. Since $\bar{\mu}_t$ is *not* stationary for plans $\tau \in \mathcal{R}_1$, the probability p_1 does not in fact remain constant (though, over the expected lifespan of a structure, it will not often change greatly). If $\bar{\mu}_t$ increases (as it will generally with a good plan), then A_h will receive fewer offspring than predicted by the calculation of p_1 at the time A_h originated. That is, the performance of A_h looks less promising relative to the current average, so trials of A_h are curtailed. If $\bar{\mu}_t$ decreases, the opposite effect occurs. Still, the expected number of offspring varies in direct relation to A_h's relative performance, so that plans in $\mathcal{R}_1$ satisfy the (informal) characterization of reproductive plans.

A slight change in the form of the algorithms in $\mathcal{R}_1$ yields a class of algorithms $\mathcal{R}_d$ wherein a time-step is a "generation" during which *each* individual $A_h(t) \in \mathcal{B}(t)$ is replaced, deterministically instead of as an expectation, by $\mu_{ht}/\bar{\mu}_t$ offspring. Thus, for $\mathcal{R}_d$, $\mathcal{B}(t + 1)$ consists of the set of *all* offspring of the individuals in $\mathcal{B}(t)$. (To keep the population level at M individuals a special kind of rounding procedure must be used to handle the cases where $\mu_{ht}/\bar{\mu}_t$ involves a fraction so that the roundings of the fractions $\mu_{ht}/\bar{\mu}_t$ sum to zero, but this need not concern us here.)

(h is the index of the individual currently producing offspring. j is a count [down] of the number of offspring produced by individual h, and b is a cumulative count of the number of offspring.)

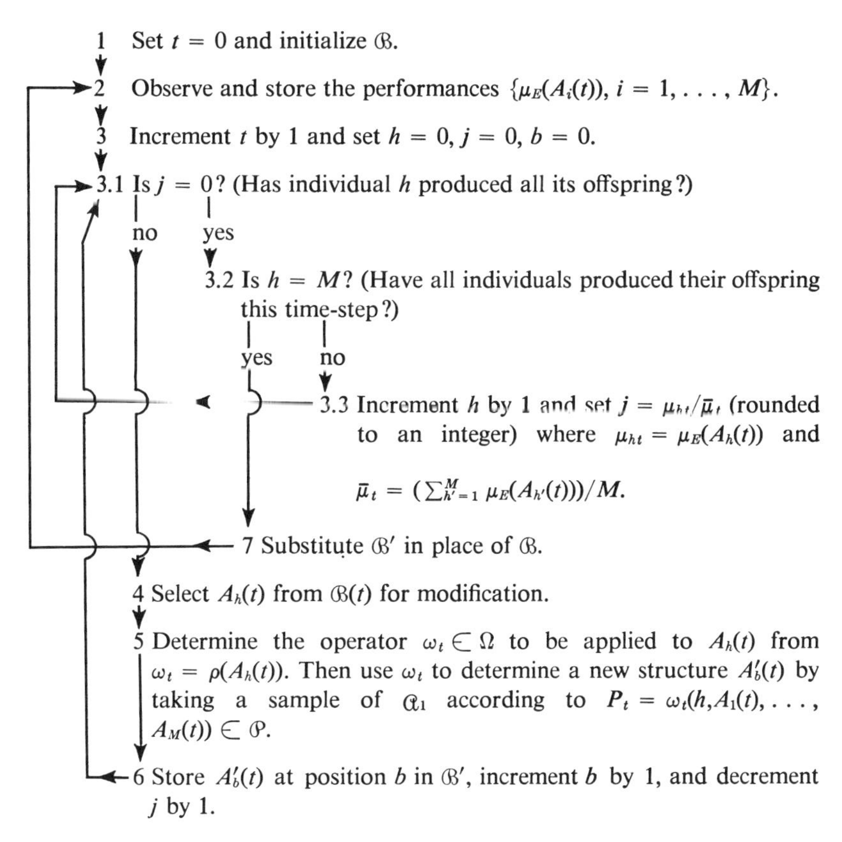

Algorithms in the class $\mathcal{R}_d$ are closer to some of the "deterministic" models of mathematical genetics. It is easier, in some respects, to interpret the role of the population $\mathcal{B}(t)$ in these plans than it is for the strictly sequential, stochastic plans in $\mathcal{R}_1$. On the other hand the algorithms in $\mathcal{R}_1$ look more like the "one-point-at-a-time" algorithms of numerical analysis and computational mathematics. Though $\mathcal{R}_1$ and $\mathcal{R}_d$ behave similarly, it is useful to have both in mind, translating from one to the other as it aids understanding.

For both types of plan the operators brought into play in step 5 are critical in determining just how past history is stored and exploited. The examination of specific operators can be expedited by subsuming $\mathcal{R}_1$ and $\mathcal{R}_d$ in a single overall diagram. Plans which satisfy this diagram and retain a recognizable variant of the "reproduction according to performance" procedures in $\mathcal{R}_1$ or $\mathcal{R}_d$ will be called plans of type $\mathcal{R}$.

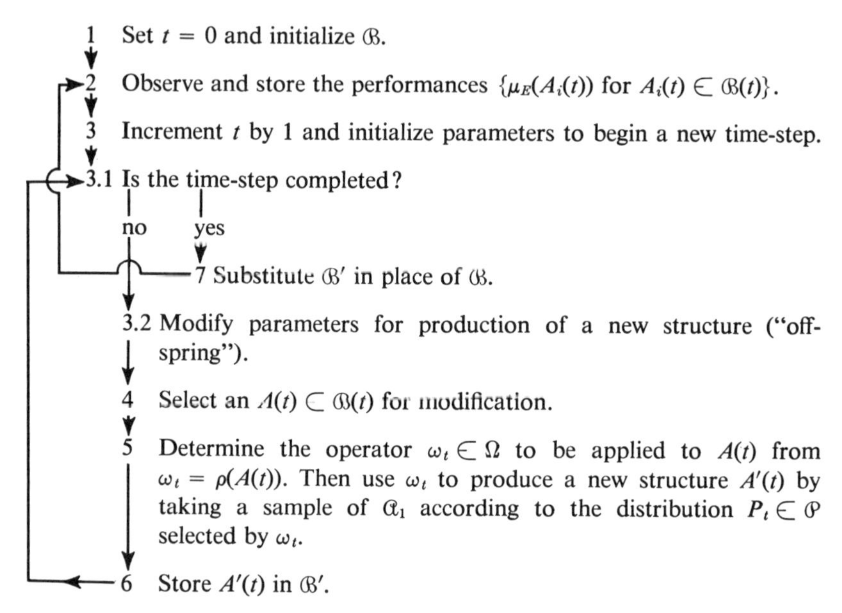

(In $\mathcal{R}_1$ steps 6 and 7 are amalgamated and the tests in 3 are unnecessary because exactly one new structure is formed per time-step.)

The next four sections will investigate the role of generalized genetic operators in plans of type $\mathcal{R}$. We will see that $\mathcal{B}(t)$ is used basically as a pool of schemata. (Recall from chapter 4 that this means that $\mathcal{B}(t)$ acts as a repository for somewhere between 2^l and $M \cdot 2^l$ schemata; i.e., it contains instances of this many distinct schemata.) Past history is recorded in terms of the ranking (number of instances) of each schema in $\mathcal{B}(t)$, much as discussed at the end of chapter 5. From this point of view *crossing-over* acts to generate new instances of schemata already in the pool while simultaneously generating (instances of) new schemata (see section 6.2). In general a total of 2^l schemata will be affected by *each* crossing-over (see Lemma 6.2.1). *Inversion* (section 6.3) affects the pool of schemata by changing the linkage (association) of alleles (attributes) defining various schemata. In combination with reproduction, the net effect is to increase the linkage of schemata of high rank (coadapted sets of alleles), making such schemata less subject to decomposition. *Mutation* (section 6.4) generally has a background role, supplying new alleles or new instances of lost alleles. All of this goes on without seriously disturbing the intrinsic rates of increase $\{\mu_\xi\}$ of most schemata instanced in $\mathcal{B}(t)$. Chapter 7 establishes the robustness and intrinsic parallelism of these type $\mathcal{R}$ plans for arbitrary string-representable domains $\mathcal{A}$.

2. GENERALIZED GENETIC OPERATORS—CROSSING-OVER

When genetic operators are used with reproductive plans we get a surprisingly sophisticated set of adaptive plans. Like the rules of a well-constructed game (chess, go, poker), genetic operators are simply defined but subtle in their consequences.

Our first objective, as with reproductive plans, will be to lift genetic operators from their specific biological context to the general $(\mathcal{J}, \mathcal{E}, \chi)$ framework. With the help of this framework we can then define and investigate rigorously two critical advantages (first discussed in chapter 4) conferred by genetic operators:

(i) intrinsic parallelism in the testing and exploitation of schemata, and
(ii) compact storage and use of the large amounts of information resulting from prior observations of schemata.

This contrasts with the common view of evolutionary processes as successive selection of the best of a sequence of variants produced by mutation—a process which we will see amounts to an enumeration of structures, with its attendant disadvantages.

The reader should be warned that the generalized operators presented in the next three sections are idealized to varying degrees. This has been done to emphasize the basic functions of the operators, at the cost of exploring the complex (and fascinating) biological mechanism underlying their execution. Even so an attempt has been made to keep the correspondence close enough to allow ready translation of the results to the original biological context.

Because it serves well as a paradigm for other genetic operators, we will look first at "crossing-over." In biological systems, crossing-over is a process yielding recombination of alleles via exchange of segments between pairs of chromosomes. We can lift this process to the level of a general operator on structures by providing the structures with representations as in chapter 4. As before, for simplicity, $\mathcal{A}$ will be taken to *be* the set of representations. Besides facilitating the generalization to arbitrary structures this emphasizes the effects of crossing-over on schemata. Crossing-over proceeds in three steps.

1. Two structures, $A = a_1a_2 \ldots a_l$ and $A' = a'_1a'_2 \ldots a'_l$, are selected (usually at random) from the current population $\mathcal{B}(t)$. (a_i and a'_i are elements of the set of attribute values V. Hence, if A_0 is the basic structure prior to representation, $\delta_i(A_0) = a_i$. Again $a_1a_2 \ldots a_l$ abbreviates $(a_1, a_2, \ldots, a_l)$, etc.)
2. A number x is selected from $\{1, 2, \ldots, l - 1\}$ (again at random).
3. Two new structures are formed from A and A' by exchanging the set of attributes to the right of position x, yielding $a_1 \ldots a_xa'_{x+1} \ldots a'_l$ and $a'_1 \ldots a'_xa_{x+1} \ldots a_l$.

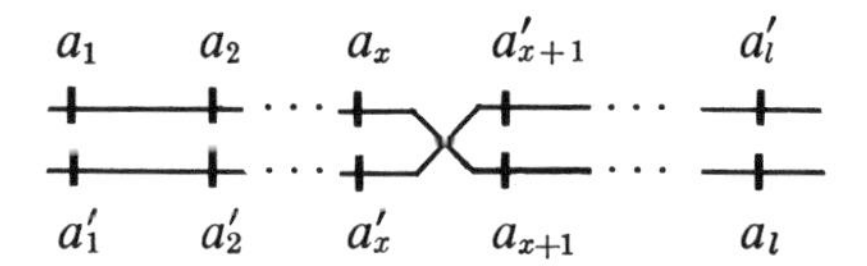

(To incorporate crossing-over directly into plans of type $\mathcal{R}$ one of the resultant structures is discarded.)

The quickest way to get a feeling for the role crossing-over plays in adaptation is to look at its effect upon schemata. To do this, consider $\mathcal{B}(t)$ as a pool of schemata (following the suggestions of chapter 4) where the number $M_\xi(t)$ of instances of ξ in $\mathcal{B}(t)$ reflects ξ's current "usefulness." The two direct effects of crossing-over on this pool are:

1. Generation of new instances of schemata already in the pool. E.g., $A = a_1a_2 \ldots a_l$ is an instance of the schema $a_1a_2 \square \ldots \square$ and, after crossing-over with $A' = a'_1a'_2 \ldots a'_l$, we have a new instance of $a_1a_2 \square \ldots \square$, namely $a_1a_2 \ldots a_xa'_{x+1} \ldots a'_l$ (assuming $a_i \neq a'_i$ for some $i \geqq x$). Each new instance of a schema ξ amounts to a new trial of the random variable corresponding to ξ. As such it increases the likelihood that the *observed* average performance $\hat{\mu}_\xi$ of the instances of ξ closely approximates the expectation μ_ξ of the random variable ξ.
2. Generation of new schemata (i.e. schemata having neither A nor A' as an instance). E.g., after the crossing-over of A with A' the schema $\square \ldots \square\ a_xa'_{x+1} \square \ldots \square$ has an instance, though neither A nor A' are instances of it (if $a_x \neq a'_x$ or $a_{x+1} \neq a'_{x+1}$). Thus $\square \ldots \square\ a_xa'_{x+1} \square \ldots \square$ will receive its first trial with the instance $a_1a_2 \ldots a_xa'_{x+1} \ldots a'_l$, unless the schema has previously been introduced to the pool from another source.

Once again $f(x)$ of Figure 10 provides a simple illustration. New instances of a schema such as $1\square\square\ldots\square$ increase confidence that the observed average $\hat{\mu}_{1\square\square\ldots\square}$ of evaluations of $f(x)$ for selected $x \in 1\square\square\ldots\square$ approaches $\mu_{1\square\square\ldots\square}$. At the same time an instance of some previously untried schema, say $110\square\ldots\square$, allows a plan of type $\mathcal{R}$ to exploit the new schema (by giving it high rank) if it is above average.

In modifying the pool of schemata, crossing-over gains tremendous power from its intrinsic parallelism. *Each* crossing-over affects great numbers of schemata, as established by the following:

LEMMA 6.2.1: *Let $A = a_1a_2\ldots a_l$ and $A' = a'_1a'_2\ldots a'_l$ differ in attribute values at x' positions to the left of $x + 1$ and x'' positions to the right of x. Then either resultant of a single crossing-over of A with A' at x will be an instance of $2^l - 2^{l-x'} - 2^{l-x''} + 2^{l-(x'+x'')}$ "new" schemata (instanced by neither A nor A'). It will also be a new instance of $2^{l-x'} + 2^{l-x''} - 2^{l-(x'+x'')}$ schemata already instanced by A or A' (assuming $x' \neq 0$ and $x'' \neq 0$).*

Proof: After crossing-over, any schema which is defined at one or more of the x' positions on the left *and* at one or more of the x'' positions on the right will have neither A nor A' as an instance. On the left there are $2^{x'} - 1$ ways of combining one or more of the x' attribute values with "$\square$'s"; similarly there are $2^{x''} - 1$ ways on the right; at the other $l - (x' + x'')$ positions either an attribute value or a "$\square$" is allowable without restriction. Thus there are $(2^{x'} - 1)(2^{x''} - 1)(2^{l-(x'+x'')}) = 2^l - 2^{l-x'} - 2^{l-x''} + 2^{l-(x'+x'')}$ "new" schemata of which the resultant is an instance.

If $x' > 0$ and $x'' > 0$ the remainder of the 2^l schemata instanced by the resultant, i.e., $2^{l-x'} + 2^{l-x''} - 2^{l-(x'+x'')}$, will have a new instance (though they are not "new" schemata) since the resultant must differ by at least one attribute value from both A and A'. Q.E.D.

In other words *each* of the 2^l schemata instanced by the resultant arises from a potentially useful manipulation of schemata already in the pool (those instanced by A and A'). Note also that, even when l is only 20, a single operation is processing over a million schemata!

We can gain additional insight concerning crossing-over by considering its effect, over an extended interval, on the whole pool of schemata in $\mathcal{B}(t)$. In the absence of reproduction and other operators, crossing-over generates a kind of diffusion from the pool to schemata not represented therein. More precisely, repeated application of crossing-over to the individuals in $\mathcal{B}(t)$ yields a "steady state" wherein, at any instant (time-step), each schema ξ has a well-defined probability of occurrence $\lambda(\xi)$. It follows that the expected interval between occurrences of ξ will be just the reciprocal $1/\lambda(\xi)$ of this probability. Thus, if the proportions of schemata in $\mathcal{B}(t)$ are not far removed from steady-state values, $1/\lambda(\xi)$ is a reasonable measure of the expected time to an occurrence of ξ. Of course, no actively adapting system (natural or artificial) following a plan of type $\mathcal{R}$ will even begin to approach the steady state. Under such a plan, the steady state is continually "modulated" by changes in the number of instances of various ξ resulting from reproduction according to $\hat{\mu}_\xi$. In effect, with reproduction added, $1/\lambda(\xi)$ is a continually changing "background" testing rate, giving at any time a rough estimate of the expected time to first occurrence of ξ. These ideas, together with values for $\lambda(\xi)$ are established rigorously by

LEMMA 6.2.2: *Repeated crossing-over (with uniform random pairing of individuals and in the absence of other operators) in a population $\mathcal{B}(t)$ yields a "steady state" (i.e., a fixed point of the stochastic transformation) in which each schema ξ occurs with probability $\lambda(\xi) = \Pi_j P({}_j\xi)$ where $P({}_j\xi)$ is the overall proportion in $\mathcal{B}(t)$ of the allele occurring at the* j*th position of ξ (if a "□" occurs at the* j*th position take $P({}_j\xi) = 1$).*

Proof: Let ${}^x\xi_1$ and ${}^x\xi_2$ be the resultants of a crossing-over of ξ_1 and ξ_2 at point x. Then a crossing-over of the resultants ${}^x\xi_1$ and ${}^x\xi_2$ at point x will bring back ξ_1 and ξ_2 (i.e., as may be determined directly from its definition, the crossover operator is self-dual).

Letting $P(\xi)$ designate the proportion of (instances of) ξ in $\mathcal{B}(t)$, we have $P(\xi_1)P(\xi_2)$ as the probability that ξ_1 will be paired with ξ_2 for crossing-over (under uniform random pairing). Thus the probability that ${}^x\xi_1$, ${}^x\xi_2$ arise from a crossing-over of ξ_1, ξ_2 at x is $P(\xi_1)P(\xi_2)P_x$, where P_x is the probability that crossover takes place at x.

Similarly the probability of a reversion (ξ_1, ξ_2 arising from ${}^x\xi_1$, ${}^x\xi_2$ by crossover at x) is $P({}^x\xi_1)P({}^x\xi_2)P_x$.

Considering only the effects of crossing-over at x on the pairs ξ_1, ξ_2 and ${}^x\xi_1$, ${}^x\xi_2$, there will be no changes in their probabilities of occurrence if

$$P(\xi_1)P(\xi_2)P_x = P({}^x\xi_1)P({}^x\xi_2)P_x.$$

If (and only if) such an equation holds for *every* x and *every* ordered quadruple $(\xi_1, \xi_2, {}^x\xi_1, {}^x\xi_2)$ will there be no change in the probability of occurrence of any schema.

To balance all of these equations simultaneously, note first that the set of alleles $\{_j\xi_1, {_j\xi_2}\}$ is identical to the set of alleles $\{^x_j\xi_1, {^x_j\xi_2}\}$ since, after crossing-over, the same alleles are still present at the jth positions (though to the right of x they will have been interchanged). Hence

$$P(_j\xi_1)P(_j\xi_2) = P(^x_j\xi_1)P(^x_j\xi_2).$$

Thus, if $P(\xi) = \Pi_j P(_j\xi)$ for each ξ, as defined in the statement of the lemma, we have for any x, ξ_1, ξ_2, ${^x\xi_1}$, ${^x\xi_2}$,

$$\begin{aligned} P(\xi_1)P(\xi_2) &= (\Pi_j P(_j\xi_1))(\Pi_j P(_j\xi_2)) = \Pi_j P(_j\xi_1)P(_j\xi_2) \\ &= \Pi_j P(^x_j\xi_1)P(^x_j\xi_2) = (\Pi_j P(^x_j\xi_1))(\Pi_j P(^x_j\xi_2)) \\ &= P(^x\xi_1)P(^x\xi_2). \end{aligned}$$

In other words, each of the equations will be balanced if the schemata occur with probabilities $\lambda(\xi) = \Pi_j P(_j\xi)$; it is also clear that any departure from these probabilities will unbalance the equations in such a way as to result in changes in some of the probabilities of occurrence. Thus, the assignment $\lambda(\xi)$ is the unique "steady state" (fixed point) of the crossover operator. Q.E.D.

We can see from the proof of this lemma that a kind of "pressure" toward the steady state

$$\Delta = P(\xi_1)P(\xi_2) - P(^x\xi_1)P(^x\xi_2)$$

can be defined for each quadruple ξ_1, ξ_2, ${^x\xi_1}$, ${^x\xi_2}$. If $\Delta \neq 0$ for any quadruple then probabilities of occurrence will start changing and there will be a diffusion toward the resultants ${^x\xi_1}$, ${^x\xi_2}$ ($\Delta > 0$) or the precursors ξ_1, ξ_2 ($\Delta < 0$). For example, if $P(\xi_1) > \lambda(\xi_1)$ while the other components remain at their steady-state values, there will be a "movement to the right"—a tendency to increase the probabilities of the result. The following heuristic argument gives some idea of the rate of approach to steady state from such departures:

A given individual has probability $2/M$ of being involved in a crossover when $\mathcal{B}(t)$ contains M individuals (since two individuals are involved in each application of the crossover operator). Thus in N trials a given individual can expect to undergo $2N/M$ crossing-overs. When N is in the vicinity of $lM/2$, where l is the length of individual representations, each individual in the population can be expected to have undergone independent crossing-over at almost every position. As a result even extreme departures from steady state should be much reduced in $lM/2$ trials.

The reduction to steady state does not, however, proceed uniformly with respect to all schemata because the crossover operator induces a *linkage* phenomenon. Simply stated, linkage arises because a schema is less likely to be affected by crossover if its defining positions are close together. In more detail, let ξ's defining positions (those not having a "$\square$") be $i_1 < i_2 < \ldots < i_h$ and let the *length* of ξ be defined as $l(\xi) = (i_h - i_1)$. Then the probability of the crossover falling somewhere in ξ, once an instance of ξ has been selected for crossing-over, is just $l(\xi)/(l - 1)$. E.g., if $A = a_1a_2a_3a_4a_5 \ldots a_l$ is selected for crossing-over, the probability of the crossover point x falling within $\xi = \square\, a_2 \square\square\, a_5 \square \ldots \square$ is $3/(l - 1)$. Clearly the smaller the length of a schema, the less likely it is to be affected by crossing-over. Thus, the smaller the length of ξ, the more slowly will a departure from $\lambda(\xi)$ be reduced.

Alleles defining a schema ξ of small length $l(\xi)$ which exhibits above-average performance will be tried ever more frequently as a unit under an adaptive plan of type $\mathcal{R}$. I.e., the alleles will be associated and tried accordingly. More modifications and tests of such schemata will be tried, and many of these trials will be of a variety of combinations with other similarly favored schemata defined at other positions. In effect such schemata serve as provisional structural elements or primitives. This observation is made precise by the following simple but important

THEOREM 6.2.3: *Consider a reproductive plan of type* $\mathcal{R}$ *using only the* simple crossover operator—*defined as a crossover operator with both precursors, and the single crossover point, determined by uniform random selection. Then the expected proportion of* each *schema represented in* $\mathcal{B}(t)$ *changes in one generation from* $P(\xi, t)$ *to*

$$P(\xi, t + 1) \geqq [1 - P_C \cdot (l(\xi)/(l - 1))(1 - P(\xi, t))](\hat{\mu}_\xi(t)/\hat{\mu}(t))P(\xi, t),$$

where P_C *is the proportion of individuals undergoing crossover during a generation and* $\hat{\mu}(t)$ *is the observed average performance of* $\mathcal{B}(t)$. *(The unit of time here—a generation—is the* expected *time for an individual to produce its offspring.)*

Proof: During one generation each individual $A \in \mathcal{B}(t)$ can be expected to produce $\mu_E(A)/\hat{\mu}(t)$ offspring under a reproductive plan of type $\mathcal{R}$. The total expected offspring of the set of instances $\mathcal{B}_\xi(t)$ of ξ in $\mathcal{B}(t)$ is thus given by

$$M'_\xi(t) = (\textstyle\sum_{A \in \mathcal{B}_\xi(t)} \mu_E(A))/\hat{\mu}(t) = \hat{\mu}_\xi(t) M_\xi(t)/\hat{\mu}(t).$$

If P_C is the proportion of $\mathcal{B}(t)$ selected to undergo crossover and $l(\xi)$ is the length of ξ, then a proportion $P_C l(\xi)/(l - 1)$ of the $M'_\xi(t)$ offspring will have a crossover falling within the defining positions of ξ. When an instance of ξ is crossed with another instance of ξ the result will also be an instance of ξ; otherwise the resultant may not be an instance of ξ. Since the probability of ξ crossing with ξ is $P(\xi, t)$ no more than a proportion $(1 - P(\xi, t))P_C l(\xi)/(l - 1)$ of the modified offspring of ξ can be expected to be instances of schemata other than ξ; the remainder $[1 - (1 - P(\xi, t))P_C l(\xi)/(l - 1)]$ will be instances of ξ.

That is,

$$\begin{aligned} P(\xi, t + 1) &= M_\xi(t + 1)/M \\ &\geqq [1 - (1 - P(\xi, t))P_C l(\xi)/(l - 1)]M_\xi'(t)/M \\ &= [1 - P_C \cdot (l(\xi)/(l - 1))(1 - P(\xi, t))](\mu_\xi(t)/\hat{\mu}(t))P(\xi, t). \end{aligned}$$

(It should be noted that crossing-over applied to precursors which are not instances of ξ may yield a resultant which is an instance of ξ. Thus $M_\xi(t + 1)$ may be enlarged, by a small amount usually, from sources outside $B_\xi(t)$; this of course only strengthens the above bound.) Q.E.D.

From this result we see that the proportion of (instances of) a schema ξ will increase as long as

$$[1 - P_C \cdot (l(\xi)/(l - 1))(1 - P(\xi, t))](\hat{\mu}_\xi(t)/\hat{\mu}(t)) \geqq 1$$

or, using the fact that $1/(1 - c) \geqq 1 + c$ for $c \leqq 1$,

$$\hat{\mu}_\xi(t) \geqq [1 + P_C \cdot (l(\xi)/(l - 1))(1 - P(\xi, t))]\hat{\mu}(t).$$

Since the worst case occurs when $P_C = 1$ (every individual in $\mathcal{B}(t)$ subjected to crossing-over) and $P(\xi, t)$ is small, we see that ξ will always increase its representation if

$$\hat{\mu}_\xi(t) \geqq [1 + (l(\xi)/(l - 1))]\hat{\mu}(t).$$

Since $1/l \leqq l(\xi)/(l - 1) \leqq 1$, short schemata need perform only slightly above average to increase, while the longest schemata (if they occur in small proportion) may have to exhibit a performance twice the population average to increase.

Theorem 6.2.3 provides the first evidence of the intrinsic parallelism of genetic plans. *Each* schema represented in the population $\mathcal{B}(t)$ increases or decreases according to the above formulation *independently of what is happening to other schemata* in the population. The proportion of each schema is essentially determined by its average performance in relation to the population average. Thus we see the evolution of a ranking of schemata based on observed performance, as suggested at the end of chapter 4 and amplified in section 5.4. Crossing-over serves this adaptive process by continually introducing new schemata for trial, while testing extant schemata in new contexts—all this without much disturbing the ranking process (except for the longer schemata). Moreover, crossing-over makes it possible for the *schemata* represented in $\mathcal{B}(t)$ to move automatically to appropriate rankings through the application of the genetic plan to individual *structures* from $\mathcal{A}$. As a result this very large number of rankings is compactly stored in a selected, relatively small population of individuals (exploiting the possibility suggested at the end of chapter 4).

By extending the pressure analogy introduced just before Theorem 6.2.3 we can gain a global view of the interaction of reproduction and crossover. Whenever some schema ξ exhibits better-than-average performance, reproduction introduces "pressures" $\Delta > 0$, disturbing the steady state which would result from the action of the crossover operator alone. The disturbances both shift the steady-state values $\lambda(\xi')$ for large numbers of schemata, because of changes in the proportions $P({}_j\xi)$ of the alleles ${}_j\xi$, $1 \leqq j \leqq l$, and also introduce local transitory departures because $P(\xi) > \lambda(\xi)$. Because all schemata are being affected simultaneously, and because reproduction affects them according to observed performance, we have a diffusion "outward" from schemata currently represented in $\mathcal{B}(t)$, a diffusion which proceeds rapidly in the vicinity of schemata exhibiting above-average performance. This is closely analogous to a gas diffusing from some central location through a medium of varying porosity, where above-average porosity is the analogue of above-average performance. The gas will exhibit a quickened rate of diffusion wherever it encounters a region of higher porosity, rapidly saturating the whole region. All the while it slowly but steadily infuses enclaves of low porosity. In effect, high porosity is exploited wherever it occurs, without prejudicing eventual penetration into regions of lower porosity. As a result the overall rate of penetration is much more determined by regions of high porosity and their proximity to each other than by average porosity.

Restated in terms of schemata, regions of higher porosity correspond to sets of schemata of above-average performance which can be produced from each other by relatively few crossovers. Thus, following the analogy, local optima in performance are thoroughly explored in an intrinsically parallel fashion. At the same time the genetic plan does not get entrapped by settling on some local optimum when further improvements are possible. Instead all observed regions of high performance are exploited without significantly slowing the overall search for better optima. Here we begin to see in a more precise context the powers of generalized genetic plans, powers first suggested in the specific context of section 1.4.

One final point: Plans of type $\mathcal{R}$ measure a schema's performance relative to the current average performance of the population. Thus, as time elapses, schemata must meet progressively higher criteria to attain (or retain) a high ranking. (This is, again, somewhat analogous to the slowed rate of occupation of a gas as it occupies successively larger volumes, higher porosity being required for the same occupation rate.) As a result, older schemata associated with local optima steadily lose ranking as better optima are located (unless the older schemata are components of the new schemata), so that capacity is not wasted on superseded regions.

The overall results of this section can be illustrated by elaborating the comment (on page 99) about $f(x)$ of Figure 10. Using 6 bits of accuracy ($l = 6$), assume $A_1 = .001100$, $A_2 = .000100$, $A_3 = .101000$, $A_4 = .110011$, and $A_5 = .011100$ have been chosen at random to form $\mathcal{B}(0)$. (The size of $\mathcal{B}(0)$, $M = 5$, is of course much too small to be realistic even for an algorithm for artificial systems, but it is adequate to illustrate the effects of crossing-over.) Looking at Figure 10 we see that $\mu_1 = f(A_1) = f(.001100) \cong \frac{1}{2}$. Similarly, $\mu_2 = f(A_2) \cong 1\frac{1}{2}$, $\mu_3 \cong 2$, $\mu_4 \cong 1\frac{3}{4}$, and $\mu_5 \cong \frac{1}{2}$. For these points $\bar{\mu} \cong \frac{5}{4}$. Accordingly A_1 will produce $\mu_1/\bar{\mu} \cong (\frac{1}{2})/(\frac{5}{4}) = \frac{2}{5}$ offspring—i.e., A_1 has about 2 chances out of 5 of being reproduced. Similarly A_2 will have $\mu_2/\bar{\mu} \cong \frac{6}{5}$ offspring; and so on. Figure 12 shows a typical outcome for a plan of type $\mathcal{R}$ using only reproduction and simple crossover on $\mathcal{B}(0)$. (Thus, for the reproduction of A_1, a trial was made of a random variable yielding 1 with probability $\frac{2}{5}$ and 0 with probability $\frac{3}{5}$—the outcome of the trial was 0.) The crossing-

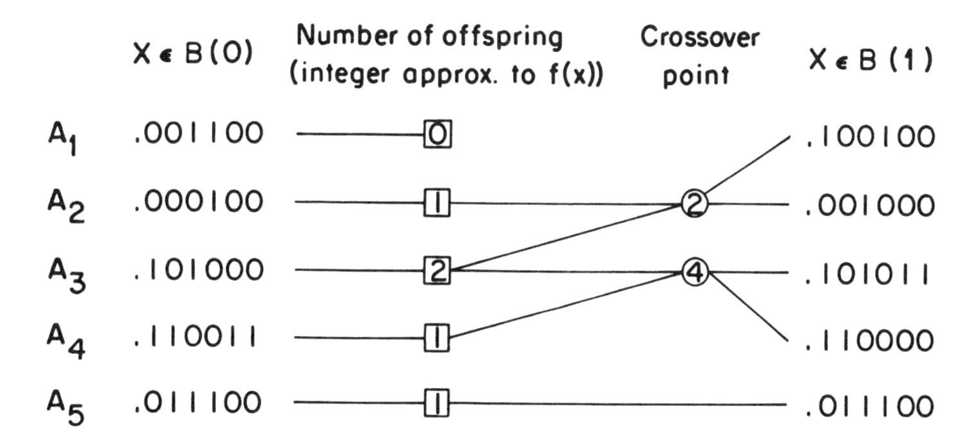

Fig. 12. Some effects of a type $\mathcal{R}$ plan on a one-dimensional function

over of A_2 with one of the replicates of A_3 at intersection ② serves both to generate another (different) instance of $1\square\square\ldots\square$, and to generate a first instance of $1\square0\square\ldots\square$. Clearly such a crossover becomes increasingly likely as instances of $1\square\square\ldots\square$ and $\square\square0\square\ldots\square$ proliferate. (Points from these schemata are likely to exhibit above-average values and hence will have more offspring on the average.) Similar effects will be happening to all other schemata instanced in $\mathcal{B}(0)$, $\mathcal{B}(1)$, etc. Figure 13 displays a more elaborate example of these effects.

3. GENERALIZED GENETIC OPERATORS—INVERSION

Crossover, by inducing a linkage between alleles, offers the possibility of an adaptable net of associations between alleles. By changing the length of a schema we modify the probability of its being affected by crossover; instances of a shorter schema are less likely to have the defining alleles separated by crossover. In consequence, under a plan of type $\mathcal{R}$, instances of the shorter schema proliferate more rapidly. The long-term effect is a selective increase in the linkage of various schemata exhibiting above-average performance. The corresponding alleles (attributes) are more frequently found in association (on the same string) in successive generations. Since schemata are defined for any string-representable domain $\mathcal{A}$, such an adaptable network of associations can be induced for any such domain by introducing an appropriate operator for changing linkage.

The linkage between the alleles defining a schema can be altered only by changing the length of the schema. That is, the positions of the alleles defining the schema (particularly the end-points) must be modifiable. However, up to this point, the functional meaning of an allele has been determined by its position. The allele a_i at the ith position of the representation of the structure A is the value $\delta_i(A)$ of the ith detector when A is its argument. Thus, if linkage is to be changed, an allele must have the same functional interpretation in any position (as is the case generally in genetics). This in turn requires a change in the method of representation.

The simplest way to change the method of representation formally is to assign each allele an index indicating the detector with which it is associated. That is, each allele is now taken to be a pair (i, a) indicating that $a = \delta_i(A)$. It follows that a structure A can be represented by any permutation of

$$((1, \delta_1(A)), (2, \delta_2(A)), \ldots, (l, \delta_l(A))).$$

For example,

$$((3, \delta_3(A)), (2, \delta_2(A)), (1, \delta_1(A)), (4, \delta_4(A)), \ldots, (l, \delta_l(a)))$$

would still represent A. Moreover, the schema $(1, \delta_1(A)) \square\square (4, \delta_4(A)) \square \ldots \square$ designates the same subset of $\mathcal{A}$ as the schema $\square\square(1, \delta_1(A))(4, \delta_4(A)) \square \ldots \square$, though the latter is more tightly linked than the former. To define this enlarged set of representations precisely, let V_i be redefined to be the set of pairs $V_i' = \{(i, v)$, for all $v \in V_i\}$ and let $\sigma^\dagger$ indicate the *set* of all permutations of the string (or l-tuple) σ. Then $\mathcal{A}^\dagger = (\Pi_{i=1}^l V_i')^\dagger$ is the enlarged set of all representations of elements in $\mathcal{A}$. The set of schemata is enlarged accordingly to $\Xi = (\Pi_{i=1}^l \{V_i' \cup \{\square\}\})^\dagger$.

The object now is to find an operator which when used with crossover and reproduction will tend to replace an above-average schema ξ with a permutation $\xi' \in \xi^\dagger$ of shorter length $l(\xi') < l(\xi)$. The genetic operator which fits this specification is *inversion*. It works by producing a crossover within a single structure as follows:

1. A structure $A = a_1 a_2 \ldots a_l$ is selected (usually at random) from the current population $\mathcal{B}(t)$ (where each a_i, $i = 1, \ldots, l$, now represents a pair $(j, v) \in V_j'$).
2. Two numbers, x_1' and x_2', are selected from $\{0, 1, 2, \ldots, l+1\}$ (again at random) and are used to define $x_1 = \min \{x_1', x_2'\}$ and $x_2 = \max \{x_1', x_2'\}$.
3. A new structure is formed from A by inverting the segment which lies to the right of position x_1 and to the left of position x_2, yielding $a_1 \ldots a_{x_1} a_{x_2-1} a_{x_2-2} \ldots a_{x_1+1} a_{x_2} \ldots a_l$.

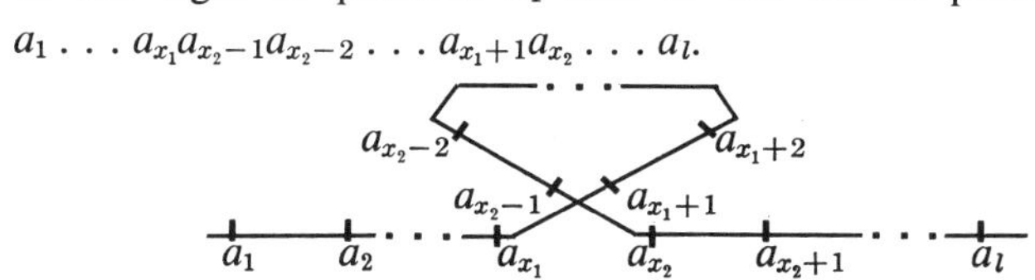

It is clear that a single inversion can bring previously widely separated alleles into close proximity, viz., a_{x_1} and a_{x_2-1} in the description. It is also clear that any possible permutation of the representation can be produced by an appropriate sequence of inversions. (More technically, the inversions $(x_1 = 0, x_2 = 2)$, $(x_1 = 1, x_2 = 3)$, $\ldots$, $(x_1 = l-1, x_2 = l+1)$ are sufficient to generate the group of all permutations of order l.) The effect of the inversion operator upon (the instances of) a schema ξ is to randomly produce permutations ξ' of ξ with varying lengths. Though inversion alters the linkage of schemata, it does *not* alter the subsets of $\mathcal{A}$ which they designate. Every permutation ξ' of ξ designates the same subset in the set of (original) structures $\mathcal{A}$ (since the same set of detector values occurs in both ξ and ξ'). The lengths of many schemata are affected simultaneously by a single inversion, so this operator too exhibits intrinsic parallelism. As with crossover, schemata of shorter lengths are less frequently affected by the inversion operator.

Let us define the *simple inversion operator* as an inversion with both the structure selected for inversion and the two points x_1 and x_2 determined by uniform random selection. To see the combined effect of simple inversion, simple crossover, and reproduction we need only refer to Theorem 6.2.3. The theorem guarantees that, if inversion has produced a permutation ξ' of ξ where $l(\xi') < l(\xi)$, then the proportion of ξ' in $\mathcal{B}(t)$ increases more rapidly than the proportion of ξ. For example, if $P_C = 1$ and $P(\xi, t)$, $P(\xi', t) \ll 1$ we can expect

$$P(\xi', t + 1) = ((l - 1 - l(\xi'))/(l - 1 - l(\xi)))(P(\xi', t)/P(\xi, t))P(\xi, t + 1)$$

since $\mu_\xi = \mu_{\xi'}$. Or, after T generations

$$P(\xi', t + T) = ((l - 1 - l(\xi'))/(l - 1 - l(\xi)))^T(P(\xi', t)/P(\xi, t))P(\xi, t + T).$$

As a result, any time inversion yields a shorter permutation ξ' of a schema ξ of above-average performance, that permutation will rapidly predominate. Because the rate of reproduction of a schema is dependent upon its length, there is a constant "pressure" toward tighter linkage of the defining alleles of schemata. Because only schemata exhibiting above-average performance occupy substantial fractions of $\mathcal{B}(t)$, the "pressure" is only important for such schemata. Inversion, by repeatedly varying the linkage, gives this pressure a chance to act.

A great many schemata are affected by each inversion, but tightly linked schemata are much less likely to be affected than loosely linked ones, so that variations are primarily in the loosely linked schemata. That is, changes in linkage are concentrated in the loosely linked (long) schemata of above-average performance, where changes are desirable. More precisely, if P_I is the proportion of the population undergoing inversion in a given generation, then the probability of a schema ξ of length $l(\xi)$ being affected is

$$2P_I \cdot (l(\xi)/(l - 1)) \cdot (1 - l(\xi)/(l - 1)) = 2P_I[l(\xi)/(l - 1) - (l(\xi)/(l - 1))^2],$$

where the second factor comes from the fact that an inversion wholly inside a schema does not affect its length. Hence, if $l(\xi) = b \cdot l(\xi') < l/4$, $b > 1$, for two schemata ξ and ξ', ξ is almost b times as likely to have its length altered.

One new restriction must be made upon the crossover operator when it is used in combination with inversion. Because of inversion, two l-tuples in $\mathcal{B}(t)$ will not always have the alleles for a given detector at the same position. Crossing-over can thus produce resultants with two (or more) alleles for a given detector, or resultants with no alleles for a given detector. For example, crossing

$$((1, a_1), (2, a_2), (3, a_3)) \quad \text{with} \quad ((1, a_1'), (3, a_3'), (2, a_2'))$$

at $x = 2$ yields $((1, a_1), (2, a_2), (2, a_2'))$ as one of the resultants. The simplest way to remedy this is to permit crossing-over only between *homologous* representations, where two representations are defined to be homologous if the detector indices (first number of each pair in the representation) are in the same order. For example, $((1, a_1), (3, a_3), (2, a_2))$ is homologous to $((1, a_1'), (3, a_3'), (2, a_2'))$, even if $a_j \neq a_j'$ for some or all j, while $((1, a_1), (2, a_2), (3, a_3))$ is not homologous to either of the foregoing. This remedy requires that the probability of inversion P_I be small so that there will exist substantial homologous subpopulations for the crossover operator to act upon. A second alternative (with a biological precedent) would be to temporarily make the second of the l-tuples chosen for crossover homologous to the first by reordering it, returning it to the population in its original order after the resultants of the crossing-over are formed. Under this alternative inversion can be unrestricted, i.e., P_I can be as large as desired.

Summing up: Inversion, in combination with reproduction and crossover, selectively increases the linkage (decreases the length) of schemata exhibiting above-average performance, and it does this in an intrinsically parallel fashion.

4. GENERALIZED GENETIC OPERATORS—MUTATION

Though mutation is one of the most familiar of the genetic operators, its role in adaptation is frequently misinterpreted. In genetics mutation is a process wherein one allele of a gene is randomly replaced by (or modified to) another to yield a new structure. Generally there is a small probability of mutation at each gene in the structure. In the formal framework this means that, each structure $A = a_1 a_2 \ldots a_l$ in the population $\mathcal{B}(t)$, is operated upon as follows:

1. The positions $x_1, x_2, \ldots, x_h$ to undergo mutation are determined (by a random process where each position has a small probability of undergoing mutation, independently of what happens at other positions).
2. A new structure $A' = a_1 \ldots a_{x_1-1} a_{x_1}' a_{x_1+1} \ldots a_{x_2-1} a_{x_2}' a_{x_2+1} \ldots a_h$ is formed where a_{x_1}' is drawn at random from the range V_i of the detector δ_i corresponding to position x_1, each element in V_i being an equilikely candidate; $a_{x_2}', \ldots, a_{x_h}'$ are determined in the same way.

If ${}^1P_M$ is the probability of mutation at each position, then the probability of h mutations in a single representation is given by the Poisson distribution with parameter ${}^1P_M$.

If successive populations are produced by mutation alone (without reproduction), the result is a random sequence of structures drawn from $\mathcal{A}$. The process is evidently enumerative (see section 1.5) since the order in which structures are generated is unaffected by the observed performances of the structures. Even a reproductive plan of type $\mathcal{R}$ using only the mutation operator is little more than an enumerative plan retaining the best structure encountered to each point in time. That is, if ${}^1P_M$ is small enough, reproduction will assure that structures with above-average performance predominate in successive generations thus retaining the better structures generated by the mutation operator. There is actually a bit of history dependence since, with ${}^1P_M$ small, the most likely structures resulting from mutation will differ by one or two alleles from the current "best" structures. Thus, the sequence of tests is not entirely random, though the dependence on observations is very unsophisticated compared to that generated by crossing-over.

Since enumerative plans are, at best, useful in very limited situations, it would seem that mutation's primary role is *not* one of generating new structures for trial—a role very efficiently filled by crossing-over. It might be objected that crossing-over cannot generate all possible combinations of alleles unless the population $\mathcal{B}(t)$ contains at least one copy of every allele. However this is not a burdensome requirement. If k is the maximum number of alleles for any detector, then as few as k strings will suffice to provide a copy of each allele. (E.g., if $V_i = \{0, 1\}$, $i = 1, \ldots, l$, then the two l-tuples $00 \ldots 0$ and $11 \ldots 1$ suffice.) There is nevertheless a difficulty which is remedied by mutation. In a population that is small relative to $\mathcal{A}$, there is always the possibility that the last copy of some allele will be eliminated during the deletion phase of a plan of type $\mathcal{R}$. Alleles which occur in structures of below-average performance are particularly susceptible; yet at some later stage these same alleles may be required in a combination necessary for further improvement. Stated another way, the lost allele may be necessary for the adaptive plan to escape a false peak. Once an allele is lost from a population, the crossover operator has no way of reintroducing it. Here, then, is a role uniquely filled by mutation, because it assures that no allele permanently disappears from the population.

Mutation introduces an additional source of loss for schemata undergoing reproduction. If the probability of mutation at each position is less than or equal to ${}^1P_M$, then a schema ξ defined on $l^0(\xi)$ positions can expect to undergo one or more mutations with probability

$$1 - (1 - {}^1P_M)^{l^0(\xi)}$$

which is approximately equal to $l^0(\xi) \cdot {}^1P_M$ when ${}^1P_M$ is small relative to $1/l$. Thus, adding mutation to the list of operators in Theorem 6.2.3, we get

COROLLARY 6.4.1: *Under a reproductive plan of type $\mathcal{R}$ using the simple crossover operator and mutation, the expected proportion of each schema represented in $\mathcal{B}(t)$ changes in one generation from $P(\xi, t)$ to*

$$P(\xi, t + 1) \geqq [1 - P_C \cdot (l(\xi)/(l - 1))(1 - P(\xi, t))] \cdot (1 - {}^1P_M)^{l^0(\xi)} \left(\frac{\hat{\mu}_\xi(t)}{\hat{\mu}(t)}\right) P(\xi, t).$$

Unlike the case for crossing-over, mutation is a constant source of loss for a schema ξ, with ${}^1P_M$ fixed, even when $P(\xi, t) = 1$. In effect it is a "disturbance" introduced to prevent entrapment on a false peak.

Summing up: Mutation is a "background" operator, assuring that the crossover operator has a full range of alleles so that the adaptive plan is not trapped on local optima. (Of course if there are many possible alleles—e.g., if we consider a great many variants of the nucleotide sequences defining a given gene—then even a large population will not contain all variants. Then mutation serves an enumerative function, producing alleles not previously tried.)

5. FURTHER INCREASES IN POWER

The next chapter will establish that the three genetic operators just described are adequate for a robust and general purpose set of adaptive plans, with one important reservation which will be discussed at the end of this section. However, there are additional operators which can make significant contributions to efficiency in more complex situations. Chief among these is the dominance-change operator which (among other things) helps to control losses resulting from mutation. Because losses resulting from mutation, for given ${}^1P_M$, do not diminish as schema ξ gains high rank, a constant "load" is placed on the adaptive plan by the random movements away from optimal configurations. For this reason it is desirable to keep the mutation rate ${}^1P_M$ as low as possible consistent with mutation's role of supplying missing alleles. In particular, if the rate of disappearance of alleles can be lowered without affecting the efficiency of the adaptive plan, then the mutation rate can be proportionally lower. Since the main cause of disappearance of alleles is sustained below-average performance, the rate of loss can be reduced by shielding such alleles from continued testing against the environment. Dominance provides just such shielding.

To introduce dominance, we must extend the method of representation once again. Pairs of alleles will be used for each detector, so that a representation involves a pair of homologous l-tuples. The object is to let some of the extra alleles be carried along with the others in an unexpressed form, forming a kind of reservoir of protected alleles. Precisely, then, the set of representations will be extended to the set of all permutations of homologous pairs $\mathcal{A}_2^\dagger = (\Pi_{i=1}^l (V_i')^2)^\dagger$. Since there is now a *pair* of alleles at each position there is no longer a direct correspondence between the detector values for a structure A and the representation of A.

Let (A', A'') be a homologous pair of l-tuples drawn from $\mathcal{A}_2^\dagger$ and let ${}_i(A', A'') \overset{df}{=} ((h, v'), (h, v''))$ where $v', v'' \subset V_h$, designate the pair of alleles occurring at the ith position of the l-tuples. The most direct way to relate this pair of l-tuples to a structure is to designate either v' or else v'' as the value of detector h, ignoring the other allele. The allele so designated will be called *dominant*, the other *recessive*. For each position i, this designation should be completely determined by information available in the pair (A', A''). Formally, for each i there should be a dominance map $d_i: \mathcal{A}_2^\dagger \to \mathcal{A}$ such that, for each $(A', A'') \in \mathcal{A}_2^\dagger$, $d_i(A', A'')$ is either the first allele or the second allele of ${}_i(A', A'')$. It should be emphasized that in this general form, the determination of the dominant allele in ${}_i(A', A'')$ may depend upon the whole context (i.e., the other alleles in (A', A'')). (This corresponds closely with Fisher's [1937, Chapter III] theory of dominance.) A simpler approach makes the determination dependent only upon the pair ${}_i(A', A'')$ itself. Thus, for each h,

$$d_h': V_h^2 \to V_h \text{ such that } d_h'(v', v'') \in \{v', v''\}$$

and

$$d_i: \mathcal{A}_2^\dagger \to \mathcal{A} \text{ such that for } {}_i(A', A'') = ((h, v'), (h, v'')),\ d_i(A', A'') = d_h(v', v'').$$

Accordingly (A', A'') represents the structure $A \in \mathcal{A}$ for which

$$\delta_h(A) = d_{i(h)}(A', A'')$$

where $i(h)$ is the index of the pair of alleles in (A', A'') for detector h.

A particularly interesting example of the simpler dominance map, useful for binary (two allele) codings (see chapter 4), can be constructed as follows. Let $V_h = \{1, 1_0, 0\}$, where 1_0 is to be recessive whenever it is paired with 0, and let the mapping $d_h: V_h^2 \to V_h$ be given by the following table:

v', v''		$d_h(v', v'')$
0	0	0
0	1_0	0
0	1	1
1_0	0	0
1_0	1_0	1
1_0	1	1
1	0	1
1	1_0	1
1	1	1

Then, for example, the representation

$$A' = ((1, 0), (3, 1_0), (2, 0), (4, 1))$$
$$A'' = ((1, 1_0), (3, 0), (2, 1), (4, 0))$$

maps to the (unpermuted) representation

$$((1, 0), (2, 1), (3, 0), (4, 1)) = 0101.$$

That is, (A', A'') represents the structure $A \in \mathcal{A}$ for which

$$\delta_1(A) = 0,\ \delta_2(A) = 1,\ \delta_3(A) = 0,\ \delta_4(A) = 1.$$

In order to examine the effect of dominance on genetic plans, the simple crossover operator must be extended to this new type of representation (inversion takes place, as before, on the individual l-tuples in the homologous pairs). To cross the homologous pair (A', A'') with the pair (A''', A''''), the procedure will be to cross A' with A''' with probability P_C, and then select one of the resultants at random. Similarly, A'' is crossed with A'''' and, again, one of the resultants is selected at random. The two selected resultants are then paired to yield one of the outcomes of the extended operation; the other two resultants are paired to yield the other outcome (if it is to be used).

To see the effect of dominance on the mutation rate, let us consider the case of two alleles v_1, v_0 at position i, where v_1 is dominant and v_0 is recessive. There are four distinct pairs of these alleles which can occur at position i in (A', A''), namely $\{(v_1, v_1), (v_1, v_0), (v_0, v_1), (v_0, v_0)\}$, and only one of these maps to v_0 under the dominance map. That is, in the pairs (v_1, v_0) and (v_0, v_1) the allele v_0 is shielded or stored without test (because the representation maps to one where only allele v_1 is present).

Stated another way, allele v_0 is only expressed or tested when it occurs in the pair (v_0, v_0). Let us assume that, on the average, the adaptive plan is to provide at least one occurrence of each allele in every T generations. That is, $P(v_0, t) \geqq 1/MT$ must be assured. In the absence of dominance (using the earlier single l-tuple representation), let the reproduction rate of v_0 (corrected for operator losses) be $(1 - \epsilon(v_0))$ exclusive of additions resulting from mutation. Then

$$P(v_0, t + 1) = (1 - \epsilon(v_0))P(V_0, t) + {}^1P_M(1 - P(v_0, t)) - {}^1P_MP(v_0, t).$$

To keep $P(v_0, t) \geqq 1/MT$ for all t, ${}^1P_M$ must be at least large enough to maintain the steady state $P(v_0, t) = P(v_0, t + 1) = 1/MT$. That is,

$$1/MT = (1 - \epsilon(v_0))/MT + {}^1P_M(1 \quad 2/MT)$$

or

$${}^1P_M = \epsilon(v_0)/((1 - 2/MT)\cdot MT).$$

If MT is at all large (as it will be for all cases of interest) this reduces to

$${}^1P_M \geqq \epsilon(v_0)/MT$$

as a close approximation to the mutation rate required without dominance. (In the extreme case that alleles v_0 are deleted whenever they are tested, ${}^1P_M = 1/MT$.) With dominance, the allele v_0 is subject to selection only when the pair (v_0, v_0) occurs. Under crossover, as extended to homologous pairs, the pair (v_0, v_0) occurs with probability $P^2(v_0, t)$. The loss from selection then is

$$2\epsilon(v_0)P^2(v_0, t)M$$

the factor 2 occurring because 2 copies of v_0 are lost each time the pair (v_0, v_0) is deleted. Again the gains from mutation are

$${}^1P_M\cdot(1 - P(v_0, t)2M - {}^1P_MP(v_0, t))2M$$

where the factor 2 occurs because the M homologous *pairs* are $2M$ l-tuples. Thus

$$P(v_0, t + 1) = P(v_0, t) - 2\epsilon(v_0)P^2(v_0, t) + {}^1P_M\cdot 2(1 - 2P(v_0, t))$$

for the homologous pairs with dominance. Setting $P(v_0, t) = P(v_0, t + 1) = 1/MT$ as before, and solving, we get

$${}^1P_M = \epsilon(v_0)/((1 - 2/MT)(MT)^2).$$

We have thus established

LEMMA 6.5.1: *To assure that, at all times, each allele a occurs with probability* $P(a, t) \geqq 1/MT$, *the mutation rate* ${}^1P_M$ *must be* $\geqq 1/MT$ *in the absence of dominance, but only* $\geqq (1/MT)^2$ *with dominance.*

For example, to sustain an average density of at least 10^{-3} for every allele, the mutation rate would have to be 10^{-3} without dominance, but only 10^{-6} with dominance.

It should be noted that, with dominance, $P(v_0, t)$ is no longer the expected *testing* rate. Although dominance allows the constant mutation load to be reduced, while maintaining a given proportion of disfavored alleles as a reserve, the testing rate of the reserved alleles is only $P^2(v_0, t)$ not $P(v_0, t)$. This reservoir is only released through a change in dominance.

Dominance change in the general case $d_i: \mathcal{A}_2^\dagger \rightarrow \mathcal{A}$, $i = 1, \ldots, l$, occurs simply through a change in context, so that dominance is directly subject to adaptation by selection of appropriate contexts. In the more restricted case $d_h: V_h^2 \rightarrow V_h$ a special operator is required. The example using $V_h = \{1, 1_0, 0\}$ will serve to illustrate the process. The basic idea will be to replace some or all occurrences of 1 by 1_0, and vice versa, in an l-tuple. Thus the previous recessives become dominant and vice versa, this change being transmitted to all progeny of the l-tuple. A simple way to do this is to designate a special inversion operator which not only inverts a segment but carries out the replacement in the inverted segment. (In genetics, there is a distant analogue in the effects produced by changes of context when a region is inverted, but it should not be taken literally.) Thus for the dominance-change inversion operator, step 3, p.107 of the inversion operator is followed by

4. In the inverted segment each occurrence of 1 is replaced by 1_0 and vice versa.

With this operator the defining alleles of an arbitrary schema can be "put in reserve" in a single operation to be "released" later, again in a single operation.

Dominance provides a reserved status not only for alleles but, more importantly, for schemata. A useful schema ξ_1 defined on many positions may be the result of an extensive search. As such it represents a considerable fragment of the adaptive plan's history, embodying important adaptations. When it is superseded by a schema ξ_2 exhibiting better performance, it is important that ξ_1 not be discarded until it is established that ξ_2 is useful over the same range of contexts as ξ_1. ξ_2's performance advantage may be temporary or restricted in some way, or ξ_1 may be useful again in some context engendered by ξ_2. In any case it is useful to retain

ξ_1 for a period comparable to the time it took to establish it. Dominance makes this possible.

Summing up: Under dominance, a given minimal rate of occurrence of alleles can be maintained with a mutation rate which is the square of the rate required in the absence of dominance. Moreover, with the dominance-change operator the combination of alleles defining a schema can be "reserved" (as recessives) or "released" (as dominants) in a single operation.

When the performance function depends upon many more or less independent factors, there is another pair of operators, *segregation* and *translocation* which can make a significant contribution to efficiency. In such situations it is useful to make provision for distinct and independent sets of associations (linkages) between genes. This again calls for an extension in the method of representation. Let each element in $\mathcal{A}$ be represented by a *set* of homologous pairs of n-tuples, and let crossover be restricted to homologous n-tuples. After two elements of $\mathcal{A}$, A and A', are chosen for crossover and after *all* homologous pairs have been crossed (as detailed under the discussion of dominance change) then from *each* pair of resultants one is chosen at random to yield the offspring's n-tuples. Each offspring thereby consists of the same number of homologous pairs of n-tuples as its progenitors. The genetic counterpart of this random selection of resultants is known as segregation. Clearly, under segregation, there is no linkage between alleles on separate nonhomologous n-tuples, while alleles on homologous n-tuples are linked as before. With this representation it is natural to provide an operator which will shift genes from one linkage set to another (so that, for example, schemata that are useful in one context of associations can be tested in another). The easiest way to accomplish this is to introduce an exceptional crossover operator, the *translocation* operator, which produces crossing-over between randomly chosen nonhomologous pairs.

Another genetic operation provides a means of adaptively modifying the effective mutation rate for different closely linked sets of alleles. The operator involved is *intrachromosomal duplication* (see Britten 1968); it acts by providing multiple copies of alleles on the same n-tuple. To interpret this operation, n-tuples with multiple copies of the alleles for a given gene must be mapped into the set of original structures. This can be done most directly by extending the concept of dominance to multiple copies of alleles. With this provision, if there are k_a copies of a given allele a, the probability of one or more mutations of allele a is k_a times greater than if there were but one copy. That is, the probability of occurrence, via mutation, of allele $a' \neq a$ is increased k_a times. Thus, increases and decreases in the number of copies of an allele have the effect of modifying the (local) mutation rate. In genetics the decreases are provided by *deletion*. The easiest generalization of these operators is an operator which doubles (or halves) the number of copies at a randomly chosen (set of adjacent) location(s).

Though the operators just described are useful, they are not necessary. Moreover they do *not* compensate the major shortcoming of genetic plans which use just the first three operators described. That shortcoming is the complete dependence of such plans upon the detectors determining the representation. If the set of detectors $\{\delta_i\}$ is inadequate, in any way, the plan must operate within that constraint. However, if the plan could add or modify detectors at need, it could circumvent the difficulty. This implies making the detectors themselves subject to adaptation. When we note that each detector can be specified by an appropriate subroutine (string of instructions) for a general purpose computer, a way of making this extension suggests itself. By keeping the number of basic instructions from which the subroutines are constructed small, we can treat them as alleles. $\mathcal{A}$ can then be extended to include all strings of basic instructions. In this way $\mathcal{A}$ contains a representation of any possible detector, set of detectors or, in fact, any effectively describable way of processing information. Moreover, under this extension, favored schemata correspond to useful coordinated sets of instructions (such as detectors). Genetic plans applied to $\mathcal{A}$, so extended, can thus develop whatever functions or representations they need. This problem and the suggested approach are complex enough to merit a chapter, chapter 8.

The Jacob-Monod (1961) "operon" model of the functioning of the chromosome has an interesting relation to the extension of $\mathcal{A}$ just suggested. In the extension, we can think of each element of $\mathcal{A}$ as a program processing inputs from the environment to produce outputs affecting that environment (cf. chapter 3.4 where transformations $\{\eta_i\}$ are the outputs). The performance of the element is thus directly determined by the relevance or fitness of the program. The "operon" model treats the chromosome as a similar information processing device. Each gene can either be active (cf. the execution of an instruction) or inactive. When active the gene is participating in the production of signals (enzymes) which modulate the ongoing activity of the cell. It thereby determines the cell's modes of action and critical aspects of its structure. The genes are collected in groups—operons—such that all genes in the group are either simultaneously active or inactive, as determined by one control gene in the group called an "operator gene" (or more recently, a "receptor gene" in Britten and Davidson 1969; see Fig. 14). The remainder of the cell is treated as the chromosome's environment. The action of the "receptor gene" is conditional upon the presence of signals (proteins) from the cell (usually through the mediation of other genes—"repressor" or "sensor" genes). In this way one operon can cause the cell to produce signals which (with controlled delays) turn on other operons. This provision for action conditional upon previous (conditional) actions gives the chromosome tremendous information-processing power. In fact, as will be shown in chapter 8, any effectively describable information-processing program can be produced in this way.

6. INTERPRETATIONS

For the geneticist, the picture of the process of adaptation which emerges from the mathematical treatment thus far exhibits certain familiar landmarks:

> Natural selection directs evolution not by accepting or rejecting mutations as they occur, but by sorting new adaptive combinations out of a gene pool of variability which has been built up through the combined action of mutation, gene recombination, and selection over many generations. For the most part Darwin's concept of *descent with modification* fits in with our modern concept of interaction between evolutionary processes, because each new adaptive combination is a modification of an adaptation to a previous environment. (p. 31)
> *Inversions* and *translocations* of chromosomal segments, when present in the heterozygous condition, can increase genetic linkage and so bind together adaptive gene combinations. . . . The importance of such increased linkage is due to the number of diverse genes which must contribute to any adaptive mechanism in a higher plant or animal. (p. 57)
>
> Stebbins in *Processes of Organic Evolution*

> Not only do we claim in this case [of inversions found in *D. pseudoobscura* and *D. persimilis*] that the precise pairing of the chromosomes in the species hybrids shows that the chromosomal material has had a common source, but we also claim that the sequence of rearrangements [produced by inversions] that occurred in the chromosome reconstructs for us the precise pattern of change that led up to and then beyond the point of speciation.
>
> Wallace in *Chromosomes, Giant Molecules, and Evolution* (p. 49)

At the same time the emphasis on gene interaction poses a series of difficult problems:

> Intricate adaptations, involving a great complexity of genetic substitutions to render them efficient would only be established, or even maintained in the species, by the agency of selective forces, the intensity of which may be thought of broadly, as proportional to their complexity.
>
> Fisher in *Evolution as a Process*, ed. Huxley et al. (p. 117)

> The interaction of genes is more and more recognized as one of the great evolutionary factors. The longer a genotype is maintained in evolution, the stronger will its developmental homeostasis, its canalizations, its system of internal feed backs become. . . . one of the real puzzles of evolution is how to break up such a perfectly co-adapted system in such a way so as not to induce extinction . . .
>
> Mayr in *Mathematical Challenges to the Neo-Darwinian Interpretation of Evolution*, ed. Moorhead & Kaplan (p. 53)

> The other and I think more interesting problem, which we have hardly begun to solve, is the question: How many changes of information are necessary to explain evolution?
>
> Waddington in *Mathematical Challenges to the Neo-Darwinian Interpretation of Evolution*, ed. Moorhead & Kaplan (p. 96)

And, even though the centennial for the *Origin of Species* has passed, speciation still lacks a general mathematical explanation. Moreover, the question of "enough time" plagues the neo-Darwinian almost as much as it did his predecessors. It is a question which weighs heavily if it is assumed that coadapted sets of alleles occur only by the spread of mutant alleles to the point that relevant combinations are likely (see Eden's [1967] comments).

In the present context each of these questions can be rephrased in terms of the processing of schemata by genetic operators. This allows us to probe the origin and development of coadapted sets of alleles much more deeply, particularly the way in which different genetic mechanisms enable exploitation of useful epistatic effects. In the next chapter, we will be able to extend Corollary 6.4.1 to demonstrate the simultaneous rapid spread of *sets* of alleles, as sets, whenever they are associated with above-average performance (because of epistasis or otherwise). Theorem 7.4 establishes the efficiency of this process for epistatic interactions of arbitrary complexity (i.e., for any fitness function $\mu:\mathcal{A} \rightarrow \mathcal{U}$, however complex). Section 7.4 gives a specific example of the process in genetic terms and exhibits a version of Fisher's (1930) theorem applicable to arbitrary coadapted sets. Finally, in section 9.3, the formalism is extended to give an approach to speciation. This extension suggests reasons for competitive exclusion within a niche, coupled with a proliferation of (hierarchically organized) species when there are many niches.

For the nongeneticist, the illustration at the end of section 6.2 should convey some of the flavor of algorithms of type $\mathcal{R}$ as optimization procedures. It is easy enough to extend that illustration to cover inversion and mutation. For example, under the revised representation of section 6.3 each bit δ is paired with a number j designating its significance (i.e. (j, δ) designates the bit $\delta \cdot 2^{-j}$). Thus bits of different orders can be set adjacent to each other in a string without changing their significance. In consequence, under the combined effect of inversion and reproduction, bits defining various regions of above-average values for $f(x)$ will be ever more tightly linked. This in turn increases the rate of exploration of intersections and refinements of these regions. Filling in the remaining details to complete the extension of the illustration is a straightforward exercise. Section 7.3 in the next chapter provides a detailed example of the response of an algorithm of type $\mathcal{R}$ to nonlinearities. Theorem 7.4 of that chapter, coupled with the comments on dimensionality in chapter 4 (p. 71) shows that, whatever the form of f (i.e., for any f mapping a bounded d dimensional space into the reals), an algorithm of type $\mathcal{R}$ optimizes expeditiously. Moreover, the algorithm does this while rapidly increasing the average value of the points it tests (though they may be scattered through many different hyperplanes), thus making the algorithm useful for "on-line" control. Sections 9.1 and 9.3 provide more detailed summaries of these advantages.

7. The Robustness of Genetic Plans

We cannot distinguish between a realistic and an unrealistic adaptive hypothesis or algorithm without a good estimate of the underlying adaptive plan's robustness—its efficiency over the range of environments it may encounter. By determining the speed and flexibility of proposed adaptive mechanisms, in the intended domain(s) of action, we gain a critical index of their adequacy. The framework of concepts and theorems has expanded now to the point that we can tackle such questions rigorously. The robustness established here is a general property holding for particular plans of type $\mathcal{R}$ in any string-represented domain $\mathcal{A}$; furthermore, the basic theorem holds for any payoff function $\mu:\mathcal{A} \rightarrow \mathcal{U}$. We can also address ourselves directly to related questions of the automatic determination, retention, and use of relevant history to increase efficiency.

1. ADAPTIVE PLANS OF TYPE $\mathcal{R}_1(P_C, P_I, {}^1P_M, \langle c_t \rangle)$

Genetic plans will be the main vehicle for this investigation, both as test cases and to illustrate formal approaches to questions of robustness. In particular, the investigation will use, as prototypes, plans of type $\mathcal{R}_1$ employing the three operators, simple crossover, simple inversion, and mutation. (To retain the one-operator format of the original specification of $\mathcal{R}_1$, the combined effect of the three operators could easily be reinterpreted as the effect of a single composite operator; for expository purposes it is easier to treat the operators individually.) The basic parameters are:

- P_C, the constant probability of applying simple crossover to a selected individual,
- P_I, the constant probability of applying simple inversion to a selected individual,
- ${}^1P_M$, the initial probability of mutation of an allele (all alternatives for the allele being equilikely outcomes),

c_t, an arbitrary sequence satisfying the conditions: (i) $0 \leqq c_t \leqq 1$, (ii) $c_t \to 0$, (iii) $\sum_t c_t \to \infty$; e.g., $c_t = (1/t)^\alpha$, $0 < \alpha \leqq 1$.

(The sequence $\langle c_t \rangle$ is included primarily for its effects when genetic plans are used as algorithms for artificial systems; it is used to drive the mutation rate to zero, while assuring that every allele is tried in all possible contexts. $\langle c_t \rangle$ is not intended to have a natural system counterpart and its effects can be ignored in that context. See below.)

1 Set $t = 0$ and initialize $\mathcal{B}$ by selecting M structures at random from $\mathcal{A}_1$ to form $\mathcal{B}(0) = (A_1(0), \ldots, A_M(0))$.

2.1 Observe and store the performances $(\mu_1(0), \ldots, \mu_M(0))$ to form $\mathcal{U}(0)$.

2.2 Calculate $\hat{\mu}(0) = \sum_{h=1}^{M} \mu_h(0)/M$.

2.3 Observe the performance $\mu_E(A'(t))$.

2.4 Update $\hat{\mu}(t)$ by calculating $\hat{\mu}(t) + \mu_E(A'(t))/M - \mu_{j(t)}(t)/M$.

2.5 Update $\mathcal{U}(t)$ by replacing $\mu_{j(t)}(t)$ with $\mu_E(A'(t))$.

3 Increment t by 1.

4 Define the random variable $Rand_t$ on $\mathcal{I}_M = \{1, \ldots, M\}$ by assigning the probability $\mu_h(t)/\hat{\mu}(t)$ to $h \in \mathcal{I}_M$. Make one trial of $Rand_t$ and designate the outcome $i(t)$.

5.1 Apply simple crossover (as defined in section 6.2 and extended in section 6.3) to $A_{i(t)}(t)$ and $A_{i'(t)}(t)$ with probability P_C, where $A_{i'(t)}(t)$ is determined by a second trial of $Rand_t$. Select one of the two resultants at random (equilikely) and designate it ${}^1A(t)$ (where the order of attributes in the resultant is that of $A_{i(t)}(t)$).

5.2 Apply simple inversion (as defined in section 6.3) with probability P_I, yielding ${}^2A(t)$.

5.3 Apply mutation (as defined in section 6.3) to ${}^2A(t)$ with probability $c_t \cdot {}^1P_M$, yielding $A'(t)$.

6.1 Assign probability $1/M$ to each $h \in \mathcal{I}_M$ and make a random trial accordingly; designate the outcome $j(t)$.

6.2 Update $\mathcal{B}(t)$ by replacing $A_{j(t)}(t)$ with $A'(t)$.

Algorithms in the subclass of $\mathcal{R}_1$ just described will be designated $\mathcal{R}_1(P_C, P_I, {}^1P_M, \langle c_t \rangle)$. The performances observed at step 2.3 will be taken to be trials of a random variable in all that follows. Thus μ_E assigns a random variable from some predetermined set $\mathcal{U}$ to each structure $\mathcal{A}_1$ (see section 2.1) and successive trials of the same structure will in general yield different performances. (For clarity, this stochastic effect is treated explicitly here, rather than using the formally equivalent approach of subsuming the effect in the stochastic action of the operators.) It will be assumed that each random variable in $\mathcal{U}$ has a well-defined mean and variance.

The study below shows that the algorithm works well and efficiently when $\mathcal{A}_1$ is small (e.g., when $\mathcal{A}_1$ has two elements as in the two-armed bandit problem) as well as when $\mathcal{A}_1$ is large. When $\mathcal{A}$ is small relative to M the genetic operators are unimportant, replication alone (step 4) being adequate to the task. However the algorithm's power is most evident when it is confronted with problems involving high dimensionality (hundreds to hundreds of thousands of attributes, as in genetics and economics) and multitudes of local optima. Computational mathematics has little to offer at present toward the solution of such problems and, when they arise in a natural context, they are consistently a barrier to understanding. For this reason it will be helpful in evaluating the algorithm to keep in mind the case where $\mathcal{A}_1$ (and $\mathcal{U}$ as well) is a very large set, finite only by virtue of a limited ability to distinguish its elements (e.g., because the detectors have a limited resolution). The ultimate finiteness of $\mathcal{A}_1$ is convenient, since then the number of attributes or detectors l can be held fixed, but it is not essential. Chapter 8 will discuss the changes required when l is an unbounded function of t.

That step 5.3 assures continued testing of all alleles in all contexts follows from

LEMMA 7.1: *Under algorithms of type $\mathcal{R}_1(P_C, P_I, {}^1P_M, \langle c_t\rangle)$ the expected number of trials of the* jth *value for the* ith *attribute (i.e., allele j of detector i), for any i and j, is infinite.*

Proof: (Essentially this proof is a specialized version of the Borel zero-one criterion.) Let $P_{ij}(t)$ be the probability of occurrence at time t of v_{ij}, the jth value for the ith attribute. Then $\sum_{t=1}^{\infty} P_{ij}(t)M$ is the expected number of occurrences of v_{ij} over the history of the system. Unless $\sum_t P_{ij}(t)M$ is infinite, v_{ij} can be expected to occur only a finite number of times. That is, unless $\sum_t P_{ij}(t)M$ is infinite, v_{ij} will at best be tested only a finite number of times in each context, and it may not be tested at all in some contexts. (Despite this a plan for which $\sum_t P_{ij}(t)M$ is large relative to the size of $\mathcal{A}_1$ may be quite interesting in practical circumstances.)

Since $\sum_t c_t \to \infty$, we have $\sum_t c_t{}^1P_M \to \infty$. But $P_{ij}(t) \geqq c_t{}^1P_M M$ for all t, whence $\sum_t P_{ij}(t)M > \sum_t c_t{}^1P_M M$. Hence $\sum_t P_{ij}(t)M$ is also infinite in the limit.

Q.E.D.

2. THE ROBUSTNESS OF PLANS $\mathcal{R}_1(P_C, P_I, {}^1P_M, \langle c_t \rangle)$

It might seem that the natural first step in establishing the robustness of an adaptive plan would be to show that it will ultimately converge to an optimal structure. However, as early as section 1.5 it was possible to make a good argument against convergence as a criterion for distinguishing useful plans. Enumerative plans converge, yet in all but the most restricted circumstances they are useless either as hypotheses or algorithms. Moreover, when data can be retained by no more than M structures and $M < |\mathcal{A}_1|$, *no* plan for searching $\mathcal{A}_1$ can yield convergence. More formally, for any $M < |\mathcal{A}_1|$, there exists $\epsilon(M) > 0$ such that as $T \rightarrow \infty$,

$$(1/T) \sum_{t=1}^{T} P(\mathcal{A}^*, t) \rightarrow 1 - \epsilon(M)$$

where $\mathcal{A}^*$ is a subset of $\mathcal{A}_1$ consisting of one or more structures with optimal mean performance (i.e., structures $A' \in \mathcal{A}^*$ such that the mean of the random variable $\mu_E(A')$ is at least as high as the mean for any $A \in \mathcal{A}_1$). This is so because for any finite sequence of trials of a suboptimal structure in $\mathcal{A}_1$, there is a non-zero probability that its *observed* average performance will exceed the *observed* performance of the optimal structure(s) (assuming overlapping distributions). Clearly, if enough of the structures being tested exhibit observed performances above that of the optimal structure(s) (again an event with a non-zero probability), the result will be the deletion of data concerning the optimal structure. Thus, unless possible convergence to a suboptimal structure is to be allowed, each structure must be repeatedly tested (infinitely often in the limit). But this repeated testing (and the law of large numbers) assures that suboptimal structures which have a finite probability of displacing an optimal structure will do so with a limiting frequency approaching that probability. Hence, for $M < |\mathcal{A}_1|$, no plan can yield $(1/T) \sum^T P(\mathcal{A}^*, t) \rightarrow 1$. At the same time, $\epsilon(M') < \epsilon(M)$ for $M' > M$ (even when $M' < |\mathcal{A}_1|$) because of two effects:

(i) the more copies there are of a suboptimal structure A in a given generation, the smaller the variance of the associated average payoff $\mu_A(t)$ (making it less likely that $\hat{\mu}_A(t) > \hat{\mu}_{A'}(t)$, $A' \in \mathcal{A}^*$),

(ii) more generations are required to displace A' in the whole population (meaning that $\hat{\mu}_A(t)$ has to exceed $\hat{\mu}_{A'}(t)$ over a longer period, a progressively less likely event).

At the cost of a small increase in the complexity of the algorithms $\mathcal{R}_1(P_C, P_I, {}^1P_M, \langle c_t \rangle)$ we can assure that they converge when $M > |\mathcal{A}_1|$ (as, for example, in the

2-armed bandit problem). The reader can learn a great deal more about convergence properties of reproductive plans from N. Martin's excellent 1973 study.

Since convergence is not a useful guide, we must turn to the stronger "minimal expected losses" criterion introduced in chapter 5. Results there (Theorems 5.1 and 5.3) indicate that the number of trials allocated to the observed best option should be an exponential function of the trials allocated to all other options. It is at once clear that enumerative plans do not fare well under this criterion. Enumeration, by definition, allocates trials in a uniform fashion, with no increase in the number of trials allocated to the observed best at any state prior to completion; accordingly, as the number of observations increases, expected losses climb precipitously in comparison to the criterion. On the other hand, plans of type $\mathcal{R}_1(P_C, P_I, {}^1P_M, \langle c_t \rangle)$ do award an exponentially increasing number of trials to the observed best, as we shall see in a moment. More importantly, plans of this type actually treat schemata from Ξ as options, rather than structures from $\mathcal{A}_1$. In doing this the plans exhibit intrinsic parallelism, effectively modifying the rank of large numbers of schemata each time a structure $A \in \mathcal{A}_1$ is tried. The effect is pronounced, even in an example as simple as that of Figure 13, which illustrates 2 generations of a small population ($M = 8$, $l = 9$) undergoing reproduction and crossover. Specifically, under plans of type $\mathcal{R}_1(P_C, P_I, {}^1P_M, \langle c_t \rangle)$ the number of instances of a schema increases (or decreases) at a rate closely related to its observed performance $\hat{\mu}_\xi(t)$ at each instant. That is, the portion $M_\xi(t)$ of *each* schema ξ represented in population $\mathcal{B}(t)$ changes simultaneously according to an equation much like that suggested at the end of chapter 5:

$$dM_\xi(t)/dt = \hat{\mu}_\xi(t)M_\xi(t).$$

The foregoing statements can be established with the help of

LEMMA 7.2: *Under a plan of type* $\mathcal{R}_1(P_C, P_I, {}^1P_M, \langle c_t \rangle)$, *given* $M_\xi(t_0)$ *instances of* ξ *in the population* $\mathcal{B}(t_0)$ *at time* t_0, *the expected number of instances of* ξ *at time* t, $M_\xi(t)$, *is bounded below by*

$$M_\xi(t_0)\Pi_{t'=t_0}^{t-1}(1 - \epsilon_\xi)\hat{\mu}_\xi(t')/\hat{\mu}(t')$$

where

$$\epsilon_\xi = (P_C + 2P_I)l(\xi)/(l - 1) + \frac{{}^1P l^0(\xi)}{M}$$

a time-invariant constant generally close to zero, depending only upon the parameters of the plan, the length $l(\xi)$ *of* ξ, *and the number* $l^0(\xi)$ *of defining positions for* ξ.

Two generations of the population $B(t) = \{A_1(t), \ldots, A_8(t)\}$:

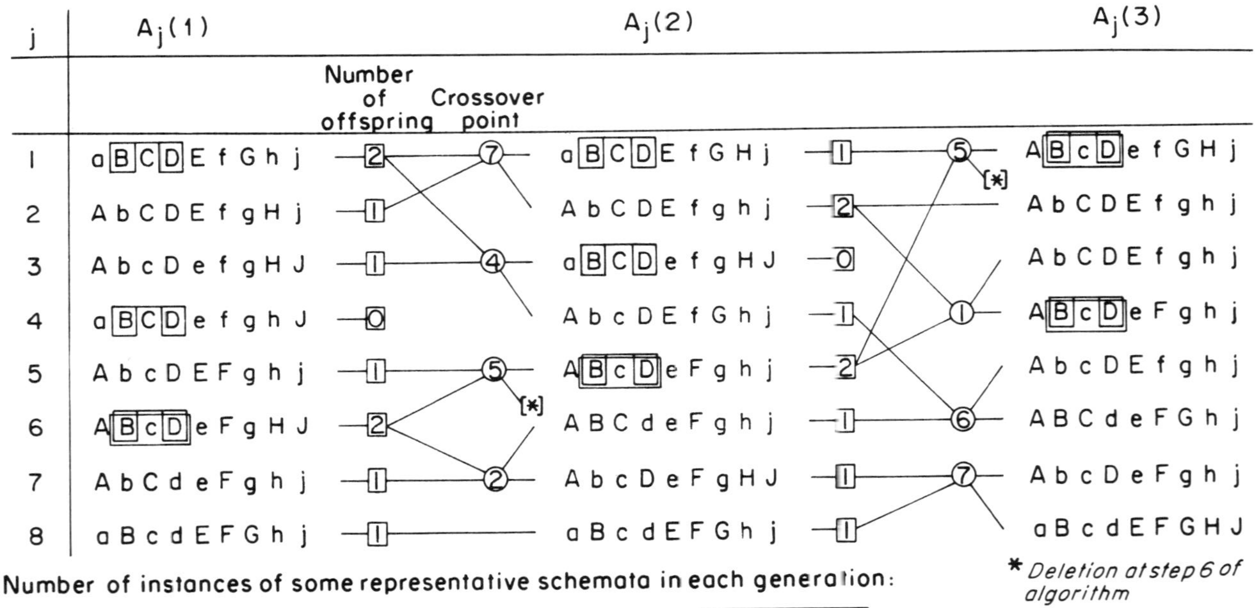

j	$A_j(1)$	Number of offspring	Crossover point	$A_j(2)$	Number of offspring	Crossover point	$A_j(3)$
1	a B C D E f G h j	2	7	a B C D E f G H j	1	5 [*]	A B c D e f G H j
2	A b C D E f g H j	1		A b C D E f g h j	2		A b C D E f g h j
3	A b c D e f g H J	1	4	a B C D e f g H J	0		A b C D E f g h j
4	a B C D e f g h J	0		A b c D E f G h j	1	1	A B c D e F g h j
5	A b c D E F g h j	1	5 [*]	A B c D e F g h j	2		A b c D E f g h j
6	A B c D e F g H J	2		A B C d e F g h j	1	6	A B C d e F G h j
7	A b C d e F g h j	1	2	A b c D e F g H J	1	7	A b c D e F g h j
8	a B c d E F G h j	1		a B c d E F G h j	1		a B c d E F G H J

* Deletion at step 6 of algorithm

Number of instances of some representative schemata in each generation:

ξ	$M_\xi(1)$	$M_\xi(2)$	$M_\xi(3)$	$\sqrt{\mu'_\xi(1)\,\mu'_\xi(2)}$
A□□□□□□□□	5	5	7	1.18
□Bc□ ... □	3	2	3	1.00
□B□D□ ... □	3	3	2	0.77
□□cD□ ... □	3	3	4	1.15
□BcD□ ... □	1	1	2	1.41
□□□dEF□□□	2	1	1	0.71
□ ... □G□□	2	3	3	1.22
□ ... □□H□	3	3	2	0.77
□ ... □GH□	0	1	2	–

$(\mu'_\xi(t) =^{df.} M_\xi(t+1)/M_\xi(t))$

Fig. 13. Example of a reproductive plan, using only crossover, applied to a small population

Proof: Using Corollary 6.4.1 in combination with the expression for the probability of a schema being affected by inversion (from section 6.3) we have, for any t_0,

$$\begin{aligned} P(\xi, t_0 + 1) &\geqq [1 - P_C\cdot(l(\xi)/(l-1))(1 - P(\xi, t_0))]\cdot[1 - 2P_I(l(\xi)/(l-1))(1 - l(\xi)/(l-1))] \\ &\quad \cdot[(1 - c_{t_0}{}^1P_M)^{l^0(\xi)}]\cdot[\hat{\mu}_\xi(t_0)/\hat{\mu}(t_0)]\cdot P(\xi, t_0) \\ &\geqq [(1 - \epsilon_\xi)\hat{\mu}_\xi(t_0)/\hat{\mu}(t_0)]P(\xi, t_0). \end{aligned}$$

Or, by iteration of the relation,

$$P(\xi, t_0 + h) \geqq P(\xi, t_0)\Pi_{t'=t_0}^{t_0+h-1}[(1 - \epsilon_\xi)\hat{\mu}_\xi(t')/\hat{\mu}(t')].$$

But the expected number of instances of ξ at time $t = t_0 + h$ is just

$$\begin{aligned} M\cdot P(\xi, t_0 + h) &\geqq M\cdot P(\xi, t_0)\Pi_{t'=t_0}^{t_0+h-1}[(1 - \epsilon_\xi)\hat{\mu}_\xi(t')/\hat{\mu}(t')] \\ &\geqq M_\xi(t_0)\Pi_{t'=t_0}^{t_0+h-1}[(1 - \epsilon_\xi)\hat{\mu}_\xi(t')/\hat{\mu}(t')]. \end{aligned}$$

Q.E.D.

Lemma 7.2, though simple, makes one very important point. Even though plans of type $\mathcal{R}_1(P_C, P_I, {}^1P_M, \langle c_t\rangle)$ try *structures* from $\mathcal{A}_1$ one at a time, it is really *schemata* which are being tested and ranked. There are somewhere between 2^l and $M2^l$ schemata with instances in $\mathcal{B}(t)$. *Each* one changes its proportion in $\mathcal{B}(t)$ at a rate largely determined by *its* observed performance, $\hat{\mu}_\xi(t)$, and is largely uninfluenced by what is happening to other schemata. This is the foundation of the intrinsic parallelism of plans of type $\mathcal{R}$.

While Lemma 7.2 is sharp enough as it stands to enable us to establish the efficiency of plans of type $\mathcal{R}_1(P_C, P_I, {}^1P_M, \langle c_t\rangle)$, some of the properties implied by the sharper inequality of the first line of its proof are also worth noting. As $P(\xi, t) \rightarrow 1$ the operator losses from crossing-over approach 0 and the first factor in brackets approaches 1. That is, as $P(\xi, t) \rightarrow 1$, the rate of change is very nearly

$$[1 - 2P_I(l(\xi)/(l-1))(1 - l(\xi)/(l-1))]\cdot[(1 - c_t{}^1P_M)^{l^0(\xi)}]\cdot[\hat{\mu}_\xi(t)/\hat{\mu}(t)] - 1.$$

Moreover, if $M_\xi(t)$ is at all large, $\hat{\mu}_\xi(t)$ will closely approximate the expected payoff $\mu_\xi(t)$ of ξ under the distribution (over $\mathcal{A}_1$ at time t) corresponding to $\mathcal{B}(t)$, because of the central limit theorem. Now recall that two schemata defined on different positions, but having identical sets of *functional* attribute pairs, designate the same subset of $\mathcal{A}_1$. Thus all permutations of ξ induced by inversion exhibit the same expected payoff $\mu_\xi(t)$ at any given time. If we treat these permutations as versions of the same schema, then inversion does not in fact result in instances of ξ being lost during the operator phase. This leaves mutation as the only important source of loss when $P(\xi, t)$ is near one. But as t advances $c_t \to 0$, so that

$$[(1 - c_t{}^1 P_M)^{l^0(\xi)}] \to 1$$

and the rate of change approaches $[\mu_\xi(t)/\mu(t)] - 1$. In particular, if some schema begins to occupy a large fraction of the population (through consistent above-average performance) its rate of increase will come very close to $[\mu_\xi(t)/\mu(t)] - 1$.

We can now go on to determine the number of trials allocated to the observed best schema as a function of the number of trials allocated to structures which are *not* instances of ξ. In this determination $n_{\xi,t_0}(t_0)$ designates the number of structures *in* $\mathcal{B}(t_0)$ which are not instances of schema ξ. N_{ξ,t_0} and n_{ξ,t_0} designate the number of trials allocated *from* t_0 *to* t to structures which are, respectively, instances of ξ and *not* instances of ξ. (That is, $N_{\xi,t_0} + n_{\xi,t_0} = (t - t_0 + 1)\cdot M$, for $t \geqq t_0$.) The logarithm of the effective payoff to ξ or *log payoff*, bounded below by $\ln[(1 - \epsilon_\xi)\hat{\mu}_\xi(t)/\hat{\mu}(t)]$, plays a direct role in

LEMMA 7.3: *If each instance of ξ gives rise, on the average, to at least one new instance of ξ in each generation over the interval (t_0, t), i.e., if $N_{\xi,t_0} \geqq (t - t_0 + 1)M_\xi(t_0 - 1)$, then the trials from t_0 onward satisfy*

$$N_{\xi,t_0} \geqq M_\xi(t_0 - 1)\exp[(Z^t_{t_0}/n_{\xi,t_0}(t_0))n_{\xi,t_0}]$$

where $Z^t_{t_0} = (1/(t - t_0 + 1))\sum_{t'=t_0}^{t}\ln[(1 - \epsilon_\xi)\hat{\mu}_\xi(t' - 1)/\hat{\mu}(t' - 1)]$ is (a lower bound on) the average log payoff over (t_0, t).

Proof:

$$\begin{aligned}
N_{\xi,t_0} &= \textstyle\sum_{t'=t_0}^{t} M_\xi(t') \\
&> M_\xi(t) \\
&\geqq M_\xi(t_0 - 1)\textstyle\prod_{t'=t_0}^{t}[(1 - \epsilon_\xi)\hat{\mu}_\xi(t' - 1)/\hat{\mu}(t' - 1)] \quad \text{using Lemma 7.2} \\
&= M_\xi(t_0 - 1)\exp[\ln \textstyle\prod_{t'=t_0}^{t}[(1 - \epsilon_\xi)\hat{\mu}_\xi(t' - 1)/\hat{\mu}(t' - 1)]] \\
&= M_\xi(t_0 - 1)\exp[\textstyle\sum_{t'=t_0}^{t}\ln[(1 - \epsilon_\xi)\hat{\mu}_\xi(t' - 1)/\hat{\mu}(t' - 1)]] \\
&= M_\xi(t_0 - 1)\exp[Z^t_{t_0}\cdot(t - t_0 + 1)]
\end{aligned}$$

However,

$$\begin{aligned}
n_{\xi,t_0}(t)/n_{\xi,t_0}(t_0) &= (M\cdot(t - t_0 + 1) - N_{\xi,t_0}(t))/(M - M_\xi(t_0)) \quad \text{by definition} \\
&\leqq (M\cdot(t - t_0 + 1) - (t - t_0 + 1)M_\xi(t_0))/(M - M_\xi(t_0)) \\
&\qquad \text{by the premise of the theorem} \\
&\leqq t - t_0 + 1
\end{aligned}$$

Substituting for $(t - t_0 + 1)$ in the previous expression we get

$$N_{\xi,t_0} \geqq M_\xi(t_0 - 1) \exp [Z_{t_0}^{\prime}(n_{\xi,t_0}/n_{\xi,t_0}(t_0))]. \qquad \text{Q.E.D.}$$

This lemma holds a fortiori for any schema ξ consistently exhibiting an effective rate of increase at least equal to 1, i.e., $\hat{\mu}_\xi(t') > \hat{\mu}(t')/(1 - \epsilon_\xi)$, over the interval (t_0, t). As noted first in sections 6.2 and 6.3, when $l(\xi)/l$ is small, ϵ_ξ will be small and the factor $1/(1 - \epsilon_\xi)$ will be very close to one. Let ξ^* denote a schema which consistently yields the best observed performance $\hat{\mu}_{\xi^*}(t')$, $t_0 \leqq t' < t$, among the schemata which persist over that interval. In all but unusual circumstances $\hat{\mu}_{\xi^*}(t')$ will exceed $\hat{\mu}(t')$ by more than the factor $1/(1 - \epsilon_\xi)$. If l is large this is the more certain since, until the adaptation is far advanced, $l(\xi)/l$ will with overwhelming probability be small—see the discussion at the end of section 6.2. Thus, for any ξ^* for which μ_{ξ^*} significantly exceeds $\hat{\mu}(t')$, $t_0 \leqq t' < t$, the number of trials N_{ξ^*,t_0} allocated to ξ^* is an exponential function of n_{ξ^*,t_0}.

(For natural systems the reproduction rate is determined by the environment—cf. fitness in genetics—hence it cannot be manipulated as a parameter of the adaptive system. However, for artificial systems this is not the case; the adaptive plan can manipulate the observed performance, as a piece of data, to produce more efficient adaptation. In particular, the reproductive step of $\mathcal{R}_1(P_C, P_I, {}^1P_M, \langle c_t \rangle)$ algorithms, step 4, can be modified to assure that the reproduction rate of ξ^* automatically exceeds $\hat{\mu}(t)/(1 - \epsilon_\xi)$.)

From all of this it is clear that, *whatever the complexity of the function* μ, plans of type $\mathcal{R}_1(P_C, P_I, {}^1P_M, \langle c_t \rangle)$ behave in a way much like that dictated by the optimal allocation criterion: the number of trials allocated to the observed best increasing as an exponential function of the total number of trials n_{ξ^*} allocated to structures which are not instances of ξ^*. However we can learn a good deal more by comparing the expected loss per trial of the genetic plans $\mathcal{R}_1(P_C, P_I, {}^1P_M, \langle c_t \rangle)$ to the loss rate under optimal allocation. Theorem 5.3 established a lower bound

$$(r - 1)b^2(\mu_{\xi^*} - \mu_2)[2 + \ln [N^2/((r - 1)^2 8\pi b^4 \ln N^2)]]$$

for the expected loss under optimal allocation, where $b = \sigma_1/(\mu_{\xi^*} - \mu_r)$. For the genetic plan, the expected loss per trial is bounded above by

$$L_P''(N) = (\mu_{\xi^*}/N)[N_{\xi^*} r' q(N_{\xi^*}, n') + (1 - r' q(N_{\xi^*}, n'))n_{\xi^*}]$$

where r' is the number of schemata which have received n' or more trials under the genetic plan, and, as in Theorem 5.3, $q(N_{\xi^*}, n')$ is the probability that a given option other than ξ^* is *observed* as best. (This expression is simply $L_{N,r}'$ from

Theorem 5.3 rewritten in the terms of the genetic plan's allocation of trials, N_{ξ^*} and n', noting that $r'q(N_{\xi^*}, n')$ is an upper bound on $q(n_1, \ldots, n_r)$.) It is critical to what follows that $r' \cdot n'$ need *not* be equal to n_{ξ^*}. As $\mathcal{B}(t)$ is transformed into $\mathcal{B}(t+1)$ by the genetic plan, *each* schema ξ having instances in $\mathcal{B}(t)$ can be expected to have $(1 - \epsilon_\xi)\mu_\xi(t)/\mu(t)$ instances in $\mathcal{B}(t+1)$. Thus, over the course of several time-steps, the number of schemata r' receiving n' trials will be much, much greater than the number of trials allocated to individuals $A \not\in \xi^*$, even when n' approaches or exceeds n_{ξ^*}. This observation, that generally $r' \cdot n' \gg n_{\xi^*}$, is an explicit consequence of the genetic plan's intrinsic parallelism.

With these observations for guidance, we can establish that the losses of genetic plans are decreased by a factor $1/(r' - 1)$ in comparison to the losses under optimal allocation. Specifically, we have

THEOREM 7.4: *If r' is the number of schemata for which*

$$n' \geqq [2Z_0^l(\xi^*)b^2/n_{\xi^*,0}(0)]n_{\xi^*,0},$$

i.e., if r' is the number of schemata for which the number of trials n' increases at least proportionally to $n_{\xi^*,0}$, then for any performance function $\mu: \mathcal{A} \rightarrow \mathcal{U}$,

$$L_p''(N)/L_r'(N) \rightarrow L < [1/(r' - 1)](\mu_{\xi^*}n_{\xi^*,0}(0)/2b^2Z_0^l(\xi^*))$$

as $N \rightarrow \infty$, where the parameters are defined as in Lemma 7.3.

Proof: Substituting the expression for N_{ξ^*} (from Lemma 7.3) and the expression for $q(N_{\xi^*}, n')$ (from the proof of Theorem 5.1) in $L_p''(N)$, and noting that $(1 - r'q(N_{\xi^*}, n'))n_{\xi^*} < n_{\xi^*}$, gives

$$L_p''(N) \lesssim (\mu_{\xi^*}/N)[(r'M_{\xi^*}(0)/\sqrt{2\pi}) \cdot \exp[(Z_0^l(\xi^*)n_{\xi^*,0}/n_{\xi^*,0}(0)) - (b^{-2}n' + \ln b^{-2}n')/2] + n_{\xi^*,0}].$$

If $b^{-2}n'/2 \geqq Z_0^l(\xi^*)n_{\xi^*,0}/n_{\xi^*,0}(0)$, it is clear that the first term (the exponential term) decreases as $n_{\xi^*,0}$ increases, but the second term, $n_{\xi^*,0}$, increases. In other words, if $n' \geqq [2Z_0^l(\xi^*)b^2/n_{\xi^*,0}(0)]n_{\xi^*,0}$, i.e., if n' increases at least proportionally with $n_{\xi^*,0}$, the expected loss per trial will soon depend almost entirely on the second term. We have already seen (in the proof of Corollary 5.2) that the same holds for the second term of the expression for expected loss under an optimal allocation of N trials. Thus, for r' and n' as specified, the ratio of *upper* bound on the reproductive plan's losses to the *lower* bound on the optimal allocation's losses approaches

$$\mu_{\xi^*}n_{\xi^*,0}/((\mu_{\xi^*} - \mu_2)(r' - 1)m^*)$$

as N increases. (This comparison yields a lower bound on the ratio since the *upper* bound in one case is being compared to the *lower* bound in the other. It can be established easily, on comparison of the respective *first* terms of the two expressions, that the condition on n' is sufficient to assure that the first term of $L_p''(N)$ is always less than the first term of $L_r'(N)$. It should be noted that the condition on n' can be made as weak as desired by simply choosing $n_{\xi^*,0}(0)$ large enough.)

To proceed, substitute the explicit expressions derived earlier for $n_{\xi^*,0}$ (Lemma 7.3) and m^* (Theorem 5.3) in the ratio $\mu_{\xi^*}n_{\xi^*,0}/((\mu_{\xi^*} - \mu_2)(r' - 1)m^*)$, yielding

$$L_p''/L_r' \rightarrow [\mu_{\xi^*}/((\mu_{\xi^*} - \mu_2)(r' - 1))](n_{\xi^*,0}(0)/Z_0^t(\xi^*)) \ln [(N - n_{\xi^*,0})/M_{\xi^*}(0)] \cdot (b^2 \ln [N^2/(8\pi b^4(r' - 1)^2 \ln N^2)])^{-1}.$$

Simplifying and deleting terms which do not affect the direction of the inequality we get

$$L_p''/L_r' \rightarrow L < [1/(r' - 1)](\mu_{\xi^*}n_{\xi^*,0}(0)/Z_0^t(\xi^*)) \cdot \ln N/[2b^2 \ln N - b^2 \ln (8\pi b^4(r' - 1)^2 \ln N^2)].$$

Or, as N grows

$$L_p''/L_r' \rightarrow L < [1/(r' - 1)](\mu_{\xi^*}n_{\xi^*,0}(0)/2b^2Z_0^t(\xi^*)). \qquad \text{Q.E.D.}$$

Thus algorithms of type $\mathcal{R}_1(P_C, P_I, {}^1P_M, \langle c_t \rangle)$ effectively exploit their intrinsic parallelism, however intricate the assignment of payoff $\mu(A)$ to structures $A \in \mathcal{A}$, reducing their losses by a factor r' in comparison to one-schema-at-a-time searches. We can get some idea of the size of r' by referring to the last few paragraphs of chapter 4. Given a representation produced by $l = 32$ detectors with $k = 2$ values ("alleles") each, $r' > 9000$ when $N = 32$ and $n' = 8$ (with all elements of $\mathcal{A}$ equally likely, i.e., $\beta_0 = \gamma_0 = 1$). This is a startling "speed-up" for a space which is, after all, small relative to the $\mathcal{A}$ spaces in, say, genetics or economics which may involve chromosomes or goods vectors with $l \gg 100$. Even small increases in N, or decreases in n' produce dramatic increases in r'; similar increases result from increases in l. Increases in l may result from representing a larger space $\mathcal{A}$, or they may be deliberately introduced for a given $\mathcal{A}$ (either by selecting $k' < k$, so that $(k')^{l'} \cong (k)^l \cong |\mathcal{A}|$ necessitates $l' > l$, or else by using additional [redundant] detectors).

To get a better picture of the implications of Lemma 7.2 and Theorem 7.4 let us look at two applications. Once again, as in chapter 1, one application is to a system which is simple and artificial, while the other is to a system which is complex and natural.

3. ROBUSTNESS VIS-À-VIS A SIMPLE ARTIFICIAL ADAPTIVE SYSTEM

The first application concerns game-playing algorithms. The game-playing illustration (section 3.3) begins by pointing out that the outcome of a 2-person game without chance moves (a strictly determined game) is fixed once each player has selected a pure strategy. Assume, for present purposes, that the opponent has adopted the best pure strategy available to him (the minimax strategy) for use in all plays of the game. Then any pure strategy selected by the adaptive plan will lead to a unique outcome and a unique payoff (again, as pointed out in section 3.3—see Figure 4). Thus, the function μ which assigns payoff to outcomes can be extended to the strategies $\mathcal{A}_1$ employed by the adaptive plan, assigning to each strategy the unique payoff it achieves against the opponent's fixed strategy. (It is helpful, though not necessary, to think of these payoffs as wide ranging—numerical equivalents of "close win," "loss by a wide margin," etc., rather than just 1, 0, -1 for "win," "draw," "loss.") The strategies available to the adaptive plan will be limited to a set of strategies fundamentally little different from the threshold pattern recognition devices of section 1.3. These strategies are based on the recognition and evaluation of positions (configurations) in the game tree and are substantially the same as those employed by Samuel in his 1959 checkers-player. Each strategy in $\mathcal{A}_1$ is defined by a linear form $\sum_{i=1}^{l} w_i\delta_i$ where: (i) $\delta_i\colon \mathcal{S} \to$ *Reals* evaluates each configuration $S \in \mathcal{S}$ for a property relevant to winning the game (e.g., in checkers, δ_1 might assign to each configuration the difference in the number of kings on each side, δ_2 might count the number of pieces advanced beyond the centerline, etc.); (ii) $w_i \in W$ weights the property according to its estimated importance in the play of the game. The linear form determines a move by assigning a rank

$$\rho(S) = \sum_{i=1}^{l} w_i\delta_i(S) \text{ to each } S \in \mathcal{S}(y), \text{ where } \mathcal{S}(y)$$

is the set of configurations legally attainable on the yth move; then that move is chosen which leads to a configuration $S^* \in \mathcal{S}(y)$ of maximal rank, i.e., $\rho(S^*) = \max_{S \in \mathcal{S}(y)}\{\rho(S)\}$.

The objective now is to find an adaptive plan which searches the set of strategies $\mathcal{A}_1$ so that performance improves rapidly. To keep the example simple only the special case of correction of weights at the end of each play of the game will be considered here. (The more complicated case, involving "predictive correction" *during* play of the game, is discussed in the latter half of section 8.4.) Because the detectors δ_i are given and fixed, the strategies in $\mathcal{A}_1$ are completely determined by the weights w_i, $i = 1, \ldots, l$, so the search is actually a search through the space of l-tuples of weights, W^l.

A typical plan for optimization in W^l adjusts the weights independently of each other (ignoring the interactions). However, in complex situations (such as playing checkers) this plan is almost certain to lead to entrapment on a false peak, or to oscillations between points distant from the optimum. Clearly such a plan is not robust. To make the reasons for this loss of robustness explicit, consider the plan $\tau'_{\mathcal{R}}$ with an initial population $\mathcal{B}(0)$ drawn from W^l, but with steps 3 and 4 of $\mathcal{R}_1(0,0,0,0)$ extended as follows.

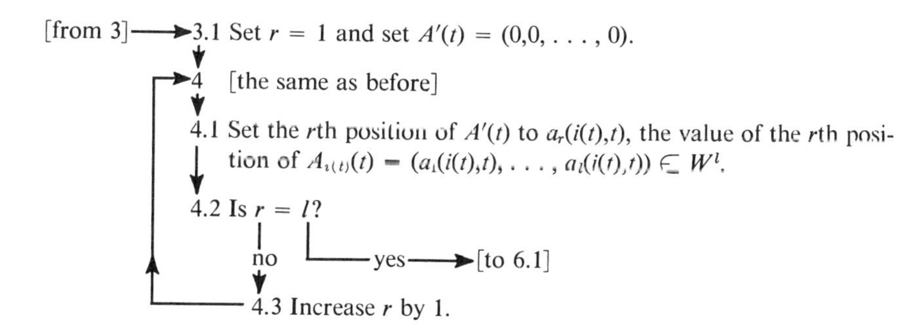

Clearly $\tau'_{\mathcal{R}}$ makes no use of the genetic operators. Over successive generations this plan has the same (stochastic) effect as repetition of the following sequence:

1. Form $\mathcal{B}'(t)$ from $\mathcal{B}(t)$ by making $\mu(A_i(t))$ copies of each element $A_i(t)$, $i = 1, \ldots, M$ in $\mathcal{B}(t)$. (Payoff $\frac{1}{2}$ yields a copy with probability $\frac{1}{2}$, so that the expected number of elements in $\mathcal{B}'(t)$ is $\sum_{i=1}^{M} \mu(A_i(t))$.)
2. All the copies of weights associated with position j of the l-tuples in $\mathcal{B}'(t)$ are collected in a single set $W_j(t)$, $j = 1, \ldots, l$. $W_j(t)$ thus, typically, contains many duplicates of each weight in W.
3. Element $A_i(t+1) = (a_1(i(t+1), t+1), \ldots, a_l(i(t+1), t+1))$, $i = 1, \ldots, M$, is formed from $\mathcal{B}'(t)$ by drawing weight $a_1(i(t+1), t+1)$ at random from set $W_1(t)$, weight $a_2(i(t+1), t+1)$ from $W_2(t)$, etc. $\mathcal{B}(t+1)$ thus consists of M l-tuples formed by M successive drawings from the l sets $W_j(t)$.
4. Return to step 1 to generate the next generation.

Because $\tau'_{\mathfrak{R}}$ makes no use of genetic operators it is a plan for adjusting weights independently. Specifically, under this procedure, the probability of occurrence of $A = a_1a_2 \ldots a_l$ at time $t + 1$ is just $\Pi_{r=1}^{l} P(a_r, t)$, where $P(a_r, t)$ is the proportion of $a_r \in W$ in $W_r(t)$. It follows at once that an arbitrary schema ξ occurs with probability $\lambda(\xi) = \Pi_j P(_j\xi)$, as would be the case under the equilibrium discussed in section 6.2. Moreover,

$$P(_j\xi, t + 1) = M_{j\xi}(t + 1)/M = (\hat{\mu}_{j\xi}(t)/\hat{\mu}(t))M_{j\xi}(t)/M$$
$$= (\hat{\mu}_{j\xi}(t)/\hat{\mu}(t))P(_j\xi, t)$$

so that

$$P(\xi, t + 1) = [\Pi_j(\hat{\mu}_{j\xi}(t)/\hat{\mu}(t))]P(\xi, t)$$

under the plan $\tau'_{\mathfrak{R}}$. Clearly the weights at distinct positions are chosen independently of each other. Hence if a *pair* of weights contributes to a better performance than could be expected from the presence of either of the two weights separately, there will be no way to preserve that observation. This can lead to quite maladaptive behavior wherein the plan ranks mediocre schemata highly and fails to exploit useful schemata. For example, consider the set of schemata defined on positions 1 and 2 when $W = \{w_1, w_2, w_3\}$. Assume that all weights are equally likely at each position (so that an instance of schema $w_2w_3\square \ldots \square$, say, occurs with probability $\frac{1}{9}$), and let the expected payoff of each schema be given by the following table:

Table 3: A Nonlinear μ_ξ on Two Positions

ξ	μ_ξ
$w_1w_1 \square \ldots \square$	0.8
$w_1w_2 \square \ldots \square$	0.3
$w_1w_3 \square \ldots \square$	1.6
$w_2w_1 \square \ldots \square$	1.1
$w_2w_2 \square \ldots \square$	1.4
$w_2w_3 \square \ldots \square$	0.8
$w_3w_1 \square \ldots \square$	1.4
$w_3w_2 \square \ldots \square$	1.3
$w_3w_3 \square \ldots \square$	0.3

Since all instances are equally likely we can calculate from this table the following expectations for single weights:

Table 4: $P(\xi)$ and μ_ξ for the One-position Schemata Implicit in Table 3

ξ	$P(\xi)$	μ_ξ
$w_1 \square\square \ldots \square$	1/3	0.9
$w_2 \square\square \ldots \square$	1/3	1.1
$w_3 \square\square \ldots \square$	1/3	1.0
$\square\, w_1 \square \ldots \square$	1/3	1.1
$\square\, w_2 \square \ldots \square$	1/3	1.0
$\square\, w_3 \square \ldots \square$	1/3	0.9

Clearly the combination w_2w_1 becomes increasingly likely under $\tau'_{\mathcal{R}}$; in fact

$$\begin{aligned} P(w_2w_1 \square \ldots \square, t+1) &= P(w_2 \square \ldots \square, t+1)\cdot P(\square\, w_1 \square \ldots \square, t+1) \\ &= [(\hat{\mu}_{w_2\square\ldots\square}(t)/\hat{\mu}(t))P(w_2 \square \ldots \square, t)] \\ &\quad \cdot[(\hat{\mu}_{\square w_1\square\ldots\square}(t)/\hat{\mu}(t))P(\square\, w_1 \square \ldots \square, t)] \\ &= 1.21\, P(w_2w_1 \square \ldots \square, t). \end{aligned}$$

On the other hand, the best combination $w_1w_3 \square \ldots \square$ by the same calculation satisfies

$$P(w_1w_3 \square \ldots \square, t+1) = 0.81\, P(w_1w_3 \square \ldots \square, t)$$

so that its probability of occurrence actually decreases. It is true that, as $w_2w_1 \square \ldots \square$ becomes more probable, the values of $\mu_{w_2\square\ldots\square}$ and $\mu_{\square w_1\square\ldots\square}$ decrease, eventually dropping below 1, but $w_1w_3 \square \ldots \square$ is still selected against, as the following table shows:

Table 5: μ_ξ for the Schemata of Table 4 when Instances Are Not Equilikely

ξ	$P(\xi)$	μ_ξ	$P(\xi)$	μ_ξ
$w_1 \square\square \ldots \square$	0.01	0.620	0.01	0.066
$w_2 \square\square \ldots \square$	0.90	0.982	0.09	1.080
$w_3 \square\square \ldots \square$	0.09	1.270	0.90	0.947
$\square\, w_1 \square \ldots \square$	0.90	0.982	0.09	1.080
$\square\, w_2 \square \ldots \square$	0.09	1.270	0.90	0.947
$\square\, w_3 \square \ldots \square$	0.01	0.620	0.01	0.066

Thus $P(w_1w_3 \square \ldots \square, t)$ steadily decreases under $\tau'_{\mathcal{R}}$, with a balance being struck among the schemata using weights $\{w_2, w_3\}$ at position 1 and weights $\{w_1, w_2\}$ at position 2. The lack of linkage between positions (or, equivalently, enforced operation at the equilibrium point $\lambda(\xi)$) destroys the robustness of $\tau'_{\mathcal{R}}$.

On the other hand the nonlinearities of μ_ξ (Table 3) have no effect on $\mathcal{R}_1(1, -, -, -)$. Lemma 7.2 makes this quite clear.

$$
\begin{aligned}
P(w_1w_3 \square \ldots \square, t+1) &= M_{w_1w_3\square\ldots\square}(t+1)/M \\
&= (1 - \epsilon_{w_1w_3\square\ldots\square})\hat{\mu}_{w_1w_3\square\ldots\square}(t)M_{w_1w_3\square\ldots\square}(t)/M \\
&= \left(1 - \frac{1}{l-1}\right)\cdot 1.6\cdot P(w_1w_3 \square \ldots \square, t),
\end{aligned}
$$

whereas $w_2w_1 \square \ldots \square$ now satisfies

$$
P(w_2w_1 \square \ldots \square, t+1) = \left(1 - \frac{1}{l-1}\right)\cdot 1.1\cdot P(w_2w_1 \square \ldots \square, t).
$$

Clearly $w_1w_3 \square \ldots \square$ quickly gains the ascendancy. Thus a plan of type $\mathcal{R}_1(1, -, -, -)$ preserves and exploits useful interactions between the weights. Moreover Lemma 7.2, in conjunction with Theorem 7.4, makes it clear that such a plan can actually exploit local optima (false peaks) to improve its interim performance on the way to a global optimum.

4. ROBUSTNESS VIS-À-VIS A COMPLEX NATURAL ADAPTIVE SYSTEM

Many points made in connection with the game-playing algorithm can be translated to the much more complex situation in genetics. We shall see that these points weigh strongly against the (still widely held) view that biological adaptation proceeds by the substitution of advantageous mutant genes under natural selection. In addition, they directly contradict the closely related view (in mathematical genetics) that alleles are replaced independently of each other, increasing or decreasing according to their individual average excesses. Rather, the results of this chapter suggest that the adaptive process works largely in terms of pools of schemata (potentially coadapted sets of genes) instead of gene pools. Because the pool of schemata corresponding to a population is so much larger than the pool of genes, selection has broader scope (some multiple of 2^l vs. $2l$, or with $k = 2$ alleles and just $l = 100$ loci, some multiple of 10^{30} vs. 200) with many more pathways to improvement, and the great advantage of intrinsic parallelism.

To translate the results on robustness to genetics, the central genetic parameter, "average excess (of fitness)," must be defined in terms of observational quantities $\hat{\mu}_\xi$. First let $\hat{\mu}'_\xi(t) \overset{\text{df.}}{=} M_\xi(t+1)/M_\xi(t)$; that is, $\hat{\mu}'_\xi(t)$ is the effective rate of increase of the schema ξ at time t. For adaptive plans of type $\mathcal{R}_1(P_C, P_I, {}^1P_M, c_t)$,

$\hat{\mu}'_\xi(t)$ is bounded below by $(1 - \epsilon_\xi)\hat{\mu}_\xi(t)$ (following Lemma 7.2). For a fraction Δt of a generation we can write

$$\begin{aligned}\Delta M_\xi(t) &= M_\xi(t + \Delta t) - M_\xi(t)\\ &= (\hat{\mu}'_\xi(t)M_\xi(t) - M_\xi(t))\Delta t\\ &= (\hat{\mu}'_\xi(t) - 1)M_\xi(t)\Delta t.\end{aligned}$$

If $M(t)$ is the size of the population at time t (allowing the overall population size to be variable for the time being), then

$$P(\xi, t) = M_\xi(t)/M(t)$$

and

$$\begin{aligned}\Delta P(\xi, t) &= \frac{M_\xi(t) + \Delta M_\xi(t)}{M(t) + \Delta M(t)} - \frac{M_\xi(t)}{M(t)}\\ &= \frac{M_\xi(t)}{M(t)} \cdot \left[\frac{1 + (\hat{\mu}'_\xi(t) - 1)\Delta t}{1 + (\hat{\mu}(t) - 1)\Delta t} - 1\right]\\ &= P(\xi, t) \cdot \frac{(\hat{\mu}'_\xi(t) - \hat{\mu}(t))\Delta t}{1 + (\hat{\mu}(t) - 1)\Delta t}\end{aligned}$$

using the fact that the population as a whole increases at a rate determined by the observed average fitness $\hat{\mu}(t)$. It follows that

$$\Delta P(\xi, t)/\Delta t = \alpha(\xi, \Delta t)P(\xi, t)$$

where $\alpha(\xi, \Delta t) =^{\text{df.}} (\hat{\mu}'_\xi(t) - \hat{\mu}(t))/(1 + (\hat{\mu}(t) - 1)\Delta t)$.

If we use a discrete time-scale $t = 1, 2, 3, \ldots$ then $\Delta t = (t + 1) - t = 1$ and

$$\alpha(\xi, 1) = (\hat{\mu}'_\xi(t) - \hat{\mu}(t))/\hat{\mu}(t).$$

If we take the limit as $\Delta t \rightarrow 0$, in effect going to a continuous time-scale, we have

$$\begin{aligned}\lim_{\Delta t \rightarrow 0}[\Delta P(\xi, t)/\Delta t] &= dP(\xi, t)/dt = \alpha(\xi, 0)P(\xi, t)\\ &= [\hat{\mu}'_\xi(t) - \hat{\mu}(t)]P(\xi, t).\end{aligned}$$

The equation $dP(\xi, t)/dt = P(\xi, t)\alpha(\xi, 0)$, when restricted to alleles (schemata defined on one position), is just Fisher's (1930) classical result, relating the change in proportion of an allele to its average excess. We see however that the equation holds for arbitrary schemata. This gives us a way of predicting the rate of increase of a set of alleles with epistatic interactions from a sample average $\hat{\mu}_\xi$ of the fitnesses of chromosomes carrying the set of alleles.

Consider, now, how such a prediction would differ from one made under the assumption of independent substitution of alleles, using the earlier example (the tables of section 7.3). In the present case the elements of W play the role of indices: w_1 at position 1 indicates the allele 1 for position 1 is present; the same w_1 at position 2 indicates the presence of allele 1 for position 2, an allele which may be quite different from the former one. Under independent selection

$$P(w_{i_1}w_{i_2}\square \ldots \square, t) = P(w_{i_1}\square\square \ldots \square, t)\cdot P(\square\, w_{i_2}\square \ldots \square, t)$$

so that

$$\frac{d}{dt}[P(w_{i_1}w_{i_2}\square \ldots \square, t)] = \frac{d}{dt}[P(w_{i_1}\square\square \ldots \square, t)\cdot P(\square\, w_{i_2}\square \ldots \square, t)]$$

$$= (P(\square\, w_{i_2}\square \ldots \square, t)\frac{d}{dt}[P(w_{i_1}\square\square \ldots, t)])$$

$$+ (P(w_{i_1}\square\square \ldots \square, t)\frac{d}{dt}[P(\square\, w_{i_2}\square \ldots \square, t)])$$

$$= P(w_{i_1}w_{i_2}\square \ldots \square, t)\cdot[\alpha(w_{i_1}\square\square \ldots \square, 0) + \alpha(\square\, w_{i_2}\square \ldots \square, 0)].$$

Thus, under *independent* selection, combinations of alleles have a rate of change which is the *sum* of their average excesses.

Reinterpreting Table 4 in terms of average excesses (noting that $\hat{\mu}(t) = 1$), we see that the rate of change of the favorable $w_1w_3\square \ldots \square$ (Table 3) is

$$-0.2\, P(w_1w_3\square \ldots \square),$$

while that of the less favorable $w_2w_1\square \ldots \square$ is $+0.2\, P(w_2w_1\square \ldots \square)$ under independent selection. Thus independent selection leads to maladaptation here.

As mentioned earlier, adaptation under independent selection amounts to adaptation under the operator equilibrium of section 6.2,

$$P(\xi, t) = \lambda(\xi, t) = \Pi_j P({}_j\xi, t).$$

This is a common assumption in mathematical genetics, but it clearly leads to maladaptations whenever

$$\alpha(\xi) \neq \textstyle\sum_j \alpha({}_j\xi).$$

The above equation for $\alpha(\xi)$ in terms of $\hat{\mu}_\xi$ shows this to be the case whenever $\mu_\xi \neq \sum_j \mu_{j\xi}$, which occurs whenever the fitness is a nonlinear function of the alleles present, i.e., whenever there is epistasis.

On the other hand, under reproductive plans of type $\mathcal{R}_1(P_C, P_I, {}^1P_M, \langle c_t \rangle)$, operator equilibrium is persistently destroyed by reproduction. In effect, useful *linkages* are preserved and nonlinearities (epistases) are exploited. Indeed, it would seem that the term "coadapted" is only reasonably used when alleles are peculiarly suited to each other, giving a performance when combined which is not simply the sum of their individual performances. Following Lemma 7.2, each coadapted set of alleles (schema) changes its proportion at a rate determined by the particular average (observed) fitness of *its* instances, *not* by the sum of the fitnesses of its component alleles.

(Because of the stochastic nature of the operators in genetic plans, each chromosome $A \in \mathcal{A}_1$ has a probability of appearing in the next generation $\mathcal{B}(t+1)$, a probability which is conditional on the elements appearing in $\mathcal{B}(t)$. If there are enough instances of ξ in $\mathcal{B}(t)$, the central limit theorem assures that $\hat{\mu}_\xi(t) \cong \mu_\xi$, where μ_ξ is the expected fitness of the coadapted set ξ under the given probability distribution over $\mathcal{A}_1$. Thus the observed rate of increase of a coadapted set of alleles ξ will closely approximate the theoretical expectation once ξ gains a foothold in the population.)

Returning to the example just above, but now for genetic ($\mathcal{R}_1$) plans, we see (from Table 3) that $w_1w_3 \square \ldots \square$ has a rate of change given by

$$+0.6 \cdot P(w_1w_3 \square \ldots \square, t),$$

while $w_2w_1 \square \ldots \square$ changes as

$$+0.1 \cdot P(w_2w_1 \square \ldots \square, t).$$

Consequently, the coadapted set of alleles with the higher average fitness quickly predominates. Thus, when epistasis is important, plans of type $\mathcal{R}_1$ (and the corresponding theorems involving schemata) provide a better hypothesis than the hypothesis of independent selection (and least mean squares estimates of the fitness of sets of alleles).

5. GENERAL CONSEQUENCES

We see from Lemmas 7.2 and 7.3 that, under a genetic plan, a schema ξ which persists in the population $\mathcal{B}(t)$ for more than a generation or two will be ranked according to its observed performance. This is accomplished in a way which satisfies the desiderata put forth at the end of chapter 5. Specifically, the proportion of ξ's instances in the population $\mathcal{B}(t)$ will grow at a rate proportional to the amount by which ξ's average performance μ_ξ exceeds the average performance μ of the whole population. At the same time the rankings are stored compactly in the way suggested at the end of chapter 4, at least 2^l schemata being ranked in a population which may consist of only a few dozen elements from $\mathcal{A}_1$. Moreover, genetic plans automatically access this information, update it, and use it to generate new structures, each of which efficiently tests large numbers of schemata.

In detail: Schemata of above-average performance are combined and tested in new contexts by crossing-over outside their defining locations. Because (the instances of) schemata increase or decrease exponentially in terms of observed performance (Lemma 7.3), the overall average performance is close to the best observed. Because a wide range of promising variants is generated and tested (section 6.2) entrapment on "false peaks" (local optima) is prevented. Even for moderate sizes of population and representation, say $M = 100$ and $l = 20$, if the initial population $\mathcal{B}(0)$ is varied, a crossover probability $P_C > \frac{1}{2}$ will make it almost certain that every structure A generated during the initial stages of adaptation is *new*. Nevertheless, this high value of P_C does *not* disturb the rankings of schemata which are consistently above average. Thus, sampling efficiency remains high, while ranking information is preserved and used. In conjunction with these processes, inversion by changes in linkage assures that schemata consistently associated with above-average performance are steadily shortened ($l(\xi)$ is decreased), thereby reducing operator losses (section 6.4 and the definition of ϵ_ξ in Lemma 7.2).

Overall, genetic plans, by simple operations on the current "data base" $\mathcal{B}(t)$, produce sophisticated, intrinsically parallel tests of the space of schemata Ξ. Large numbers of local optima, instead of diverting the plan from further improvement, are exploited to improve performance on an interim basis while the search for more global optima goes on. High dimensionality (such as a multitude of factors affecting fitness or play of a game) creates no difficulties for genetic plans, in contrast to its effect on classical procedures, because of the intrinsic parallelism (the r' factor of Theorem 7.4).

8. Adaptation of Codings and Representations

To this point the major limitation of genetic plans has been their dependence upon the fixed representation of the structures $\mathcal{A}$. The object of the present chapter is to show how to relax this limitation by subjecting the representation itself to adaptation. This will be approached by reconsidering representation via detectors $\{\delta_i:\mathcal{A} \to V_i, i = 1, \ldots, l\}$ (chapter 4) in the light of the comment that detectors can be looked upon as algorithms for assigning attributes (section 3.4). Since algorithms can be presented as strings of instructions, the possibility opens of treating them by genetic plans, much as the strings of attributes are treated. (The mode of action of the genetic operators, of course, puts some unique requirements on the form and interaction of the instructions.) Actually, with a set of instructions of adequate power, we can go much further. We can define structures capable of achieving any effectively describable behavior vis-à-vis the environment. We can do this by setting up algorithms which act conditionally in terms of environmental and internal conditions. In particular, the predictive modeling technique of sections 3.4 and 3.5 can be implemented and subjected to adaptation. The Jacob-Monod "operon-operator" model (see the end of chapter 6) is suggestive in this respect, and we'll look at it more closely after the question of a "language of algorithms" (the instructions and their grammar) has been considered.

1. FIXED REPRESENTATION

Before proceeding to a "language" suited to the modification of representations it is worth looking at just how flexible a *fixed* representation can be. That a fixed representation has limitations is clear from the fact that only a limited number of subsets of $\mathcal{A}$ can be represented or defined in terms of schemata based on that representation. If $\mathcal{A}$ is a set of structures uniquely represented by l detectors, each taking on k values, then of the 2^{kl} distinct subsets of $\mathcal{A}$ only $(k + 1)^l$ can be defined by schemata. However, the question is not so much one of defining all possible subsets, as it is one of defining enough "enriched" subsets, where an "enriched" subset is one which contains an above-average number of high-performance structures.

It is instructive, then, to determine how many schemata (on the given representation) are "enriched" in the foregoing sense. Let $\mathcal{A}$ contain x structures which are of interest at time t (because their performance exceeds the average by some specified amount). If the attributes are randomly distributed over structures, determination of "enrichment" is a straightforward combinatoric exercise. More precisely, let each δ_i be a pseudo-random function and let $V_i = \{0, 1\}, i = 1, \ldots, l$, so that a given structure $A \in \mathcal{A}$ has property i (i.e., $\delta_i(A) = 1$) with probability $\frac{1}{2}$. Under this arrangement peculiarities of the payoff function cannot bias concentration of exceptional structures in relation to schemata.

Now, two exceptional structures can belong to the same schema only if they are assigned the same attributes on the same defining positions. If there are h defining positions this occurs with probability $(\frac{1}{2})^h$. For j exceptional structures, instead of 2, the probability is $(1/2^{j-1})^h$. Since there are $\binom{l}{h}$ ways of choosing h out of l detectors, and $\binom{x}{j}$ ways of choosing j out of x exceptional structures, the expected number of schemata defined on h positions and containing exactly j exceptional structures is

$$(1/2^{j-1})^h \binom{l}{h} \binom{x}{j}.$$

For example, with $l = 40$ and $x = 10^5$ (so that the density of exceptional structures is $x/2^l = 10^5/2^{40} \cong 10^{-7}$), $h = 20$, and $j = 10$, this comes to

$$(1/2^9)^{20}(40!/(20!20!))(10^5!/(99{,}990!10!)) \cong 3.$$

Noting that a schema defined on 20 positions out of 40 has $2^{20} = 10^6$ instances, we see that the 10 exceptional structures occur with density 10^{-5}, an "enrichment" factor of 100. A few additional calculations show that in excess of 20 schemata defined on 20 positions contain 10 or more exceptional structures.

For given h and j, the "enrichment" factor rises steeply as l increases. On the other hand an increase in x (corresponding to an extension of interest to structures with performances not so far above the average) acts most directly on the expected number of schemata containing j structures. With an adequate number of pseudo-random functions as detectors (and a procedure for assuring that every combination of attributes designates a testable structure), the adaptive plan will have adequate grist for its mill. Stated another way, even when there can be no correlation between attributes and performance, the set of schemata cuts through the space of structures in enough ways to provide a variety of "enriched" subsets. Intrinsic parallelism assures that these subsets will be rapidly explored and exploited.

2. THE "BROADCAST LANGUAGE"

Though the foregoing is encouraging as to the range of partitions offered by a given set of schemata, something more is desirable when long-term adaptation is involved. First of all, when the payoff function is very complex, it is desirable to adapt the representation so that correlations between attributes and performance are generated. Both higher proportions of "enriched" schemata and higher "enrichment" factors result. It is still more important, when the environment provides signals in addition to payoff, that the adaptive plan be able to model the environment by means of appropriate structures (the $\mathfrak{M}$ component of $\mathfrak{A} = \mathfrak{A}_1 \times \mathfrak{M}$ in section 2.1). In this way large (non-payoff) information flows from the environment can be used to improve performance. As suggested in sections 3.4 and 3.5, by a process of generating predictions with the model, observing subsequent outcomes, and then compensating the model for false predictions, adaptation can take place even when payoff is a rare occurrence.

To provide these possibilities, the set of representations and models available to the plan must be defined. Further flexibility results if provision is made, within the same framework, for defining operators useful in modifying representations and models. A natural way to do this is to provide a "language" tailored to the precise specification of the representations and operators—a language which can be employed by the adaptive plan. Some earlier observations suggest additional, desirable properties of this language:

1. It should be convenient to present the representations, models, operators, etc. as strings so that schemata and generalized genetic operators can be defined for these extensions.
2. The functional "units" (cf. detectors, etc.) should have the same interpretation (function) regardless of their positioning within a string, so that advantage can be taken of the associations provided by positional proximity (section 6.3).
3. The number of alternatives at each position in a string should be small so that a richer set of schemata is provided for a given size of $\mathfrak{A}$ (see the comments in the middle of chapter 4).

4. "Completeness," in the sense of being able to define within the language all effective representations and operators, should be provided so that the language itself places no long-term limits on the adaptive plan.

What follows is an outline of one "language" satisfying these conditions. It has the additional property (which will be discussed after the presentation) of offering straightforward representations of several models of natural systems, including operon-operator models, cell-assembly models (section 3.6) and various physical signaling and radiation models.

The basic units of the language are *broadcast units*. Each broadcast unit can be thought of as broadcasting an output signal "to whom it may concern" whenever it detects certain other signals in its environment. For example, a given unit, upon detecting the presence of signals I and I' (perhaps broadcast by other units), would broadcast signal I''. Some broadcast units actually process signals so that the signal broadcast is some modification of the signals detected. In keeping with suggestion (1), broadcast units are specified by strings of symbols. A set of broadcast units, usually combined in a string, will constitute a *device* or structure (an element of $\mathcal{A}$). Some broadcast units broadcast strings which can be interpreted as (new) broadcast units; broadcast units can also detect the presence of other broadcast units (treating them as signals). Thus, given broadcast units can modify and create others—they serve as operators on $\mathcal{A}$.

The language's ten symbols $\Lambda = \{0, 1, *, :, \Diamond, \bigtriangledown, \blacktriangledown, \triangle, p, '\}$, along with informal descriptions of intended usage, follow. (Exact interpretations for strings of the symbols follow the listing.)

0, 1 These two symbols constitute the basic alphabet for specifying signals. Thus 01011 is a signal which, within the language, has no other meaning than its ability to activate certain broadcast units. Generally a string such as 01011 will be interpreted as the *name* of a particular signal (e.g. the binary encoding of a frequency or amino acid sequence).

* * indicates that the following string of symbols (up to the next occurrence of a *, if any) is to be interpreted as an active broadcast unit.

: This symbol is the basic punctuation mark, used in separating the arguments of a broadcast unit.

For example, *1100:11 designates a broadcast unit which will broadcast the signal 11 one unit of time after the signal 1100 is detected in the unit's environment (see the intended interpretations for strings below).

◇ When this symbol occurs in the argument of a broadcast unit it indicates a "don't care" condition. I.e. any symbol can occur at that particular position of a signal without affecting its acceptance or rejection by the broadcast unit. If the symbol ◇ occurs at the last position of an argument it indicates that any terminal string (suffix) may occur from that point onward without affecting acceptance or rejection.

For example, *1◇00:11 will broadcast 11 if it detects either the signal 1100 or the signal 1000 (or in fact a signal with any of the other symbols at the second position); *100◇:11 will broadcast 11 if it detects any string having the prefix 100, such as 1001 or 10010110 or even 100.

▽ ▽ designates an arbitrary initial or terminal string of symbols (an arbitrary prefix or suffix) when used in the arguments of a given broadcast unit. This symbol gives the unit string-processing capability.

For example, *11▽:▽ will broadcast the suffix of any signal having the symbols 11 as prefix. Thus if 1100 is detected, the signal 00 will be broadcast, whereas if the signal 11010 is detected, the signal 010 will be broadcast. (The resolution of conflicts, where more than one signal satisfies the input condition, is detailed below.) All occurrences of ▽ within a *given* broadcast unit designate the same substring, but occurrences in *different* broadcast units are independent of each other.

▼ A second symbol used in the same manner as ▽. It makes concatenation of inputs possible (see below).

△ This symbol serves much as ▽ and ▼ but designates an arbitrary *single* symbol in the arguments of a given broadcast unit.

For example, *11△0:1△ broadcasts 10 if 1100 is detected, or 11 if 1110 is detected.

p When *p* occurs as the first symbol of a string it designates a string which persists through time (until deleted), even though it is not a broadcast unit.

' This symbol is used to quote symbols in the arguments of a broadcast unit.

For example, *11'◇: 10 broadcasts 10 only if the (unique) string 11◇ is detected (i.e., the symbol ◇ occurs literally at the third position); without the quote the unit would broadcast 10 whenever any 3 symbol string with the prefix 11 is detected.

The interpretations of the various strings from Λ^*, the set of strings over Λ, along with the conventions for resolving conflicts, follow.

Let I be an arbitrary string from Λ^*. In I a symbol is said to be quoted if the symbol ′ occurs at its immediate left. I is parsed into broadcast units as follows: The first broadcast unit is designated by the segment from the leftmost unquoted * to (but not including) the next unquoted * on the right (if any). (Any prefix to the left of the leftmost unquoted * is ignored.) The second, third, etc., broadcast units are obtained by repeating this procedure for each successive unquoted * from the left. If I contains no unquoted * s it designates the *null* unit, i.e., it does not broadcast a signal under any condition. Thus

$$p10{*}11'{*}\triangle0{:}1\triangle{*}{:}11\nabla{:}11\nabla$$

designates two broadcast units, namely

$${*}11'{*}\triangle0{:}1\triangle \quad \text{and} \quad {*}{:}11\nabla{:}11\nabla.$$

There are four types of broadcast unit (other than the null unit). To determine the type of a broadcast unit from its designation, first determine if there are three or more (unquoted) : to the right of the *. If so ignore the third : and *everything to the right of it.* The remaining substring, which has a * at the initial positions and at most two : s elsewhere, designates one of the four types if it has one of the following four organizations.

1. ${*}I_1{:}I_2$
2. ${*}{:}I_1{:}I_2$
3. ${*}I_1{:}{:}I_2$
4. ${*}I_1{:}I_2{:}I_3$

where I_1, I_2, and I_3 are arbitrary non-null strings from Λ^* except that they contain neither unquoted * s nor unquoted : s. If the substring does not have one of these organizations it designates the null unit. The four basic types have the following functions (subject to the conventions for eliminating ambiguities, which follow).

1. ${*}I_1{:}I_2$ — If a signal of type I_1 is present at time t, then the signal I_2 is broadcast at time $t + 1$.
2. ${*}{:}I_1{:}I_2$ — If there is *no* signal of type I_1 present at time t, then the signal I_2 is broadcast at time $t + 1$.
3. ${*}I_1{:}{:}I_2$ — If a signal of type I_1 is present at time t, then a persistent string of type I_2 (if any exists) is deleted at the *end* of time t.

4. $*I_1:I_2:I_3$ — If a signal of type I_1 *and* a signal of type I_2 are both present at time t, then the signal I_3 is broadcast *at the same time t* unless I_3 contains unquoted occurrences of the symbols $\{\triangledown, \blacktriangledown, \vartriangle\}$ or singly quoted occurrences of *, in which case I_3 is broadcast at time $t + 1$.

When the final string of any of these units (I_2 for (1), (2), and (3), I_3 for (4)) is interpreted for broadcast, one quote is stripped from each quoted symbol.

The concept of the *state* of a (finite) collection of broadcast units facilitates discussion of potential ambiguities in the actions and interactions of the four types of unit. This state at time t is, quite simply, the set of *all* signals present at time t, including the strings defining devices in the collection, the signals generated by those devices, and the signals generated in the environment of the collection (input signals). Thus the *initial state* is the set of strings used to specify the initial collection of units, together with all signals present initially. If we look again at the definition of type 4 broadcast units, we see that they may actually use signals in the current state to contribute additional signals to the current state (i.e., they can act with negligible delay much as the switching elements of computer theory). For example, given the broadcast units

$$*11\triangledown:0\blacktriangledown:11\triangledown\blacktriangledown$$
$$*100:100:000$$

with environmental (input) signals $\{100, 110\}$ at $t = 0$ and $\{100\}$ at $t = 1$, the state $S(0)$ at $t = 0$ is

$$S(0) = \{*11\triangledown:0\blacktriangledown:11\triangledown\blacktriangledown,\ *100:100:000,\ 100,\ 110,\ 000\},$$

and at $t = 1$ it is

$$S(1) = \{*11\triangledown:0\blacktriangledown:11\triangledown\blacktriangledown,\ *100:100:000,\ 100,\ 000,\ 11000\}.$$

The latter signal in $S(1)$, 11000, occurs because the unit $*11\triangledown:0\blacktriangledown:11\triangledown\blacktriangledown$ receives both the signal 110 and the signal 000 at $t = 0$, so that $\triangledown = 0$ and $\blacktriangledown = 00$, and hence the output 11000 occurs at time $t = 1$. A little thought shows that the instantaneous action of type 4 units does not interfere with the determination of a unique state at each time since type 4 units can add at most a finite number of signals to the current state.

Since the symbols $\triangledown$ and $\blacktriangledown$ are meant only to designate initial or terminal strings their placement within arguments of a broadcast unit can be critical to unambiguous interpretation. For types 1 through 4, if I_1 contains exactly one unquoted occurrence of a symbol from the set {▽, ▼}, then that symbol must occur in either the first or last position to be interpreted; otherwise ▽ or ▼ is treated simply as a null symbol without function or interpretation. If I_1 contains more than one unquoted occurrence of symbols from the set {▽, ▼} then only the leftmost is operative and then only if it occupies the first position. For type 4 the same convention applies to I_2, with the additional stipulation that, if the operative symbol is the same in both I_1 and I_2, then only the occurrence in I_1 is interpreted. Similarly, for types 1 through 4, only the leftmost occurrence of an unquoted △ is operative. Moreover, if ▽, ▼, or △ occur unquoted in the output signal of the broadcast unit without interpretable occurrences in the arguments, they are once again treated as null symbols (and are not broadcast). Thus the broadcast unit *▽11▽△0:11△ "looks for" any signal with a 4 symbol suffix beginning with 11 and ending in 0; for example, the signal 001110 would yield the output 111 one time-step later.

The final source of ambiguity arises when two or more signals satisfy the same argument of a given string-processing broadcast unit. For example, when the state is

$$S(t) = \{*11\triangledown:\triangledown,\ 111,\ 1100\}$$

the broadcast unit *11▽:▽ could process either 11 or 1100, producing either the output 1 or else the output 00. This difficulty is resolved by having the unit select one of the two signals at random. That is, if there are c signals satisfying a given argument at time t, then each is assigned a probability $1/c$ and one is chosen at random under this distribution. This method of resolving the difficulty extends the power of the language, allowing the representation of random processes.

3. USAGE

The following examples exhibit typical constructions and operations within the "broadcast language":

1. The object is to produce the concatenation of two arbitrary persistent strings uniquely identified by the prefixes I_1 and I_2 respectively. In so doing the prefixes should be dropped and the result should be identified by the new prefix I_3. This is accomplished by the broadcast unit

$$*pI_1\triangledown:pI_2\blacktriangledown:pI_3\triangledown\blacktriangledown.$$

2. The object is to generate a sample of the random variable defined by assigning probability $1/n$ to each of the numbers $\{1, 2, \ldots, n\}$. To do this each

string in a set of n persistent strings, say the binary representations of the numbers 1 through n, is prefixed by the same string, say I, which uniquely indicates that the strings are to serve as the data base for the random number generator. When the state $S(t)$ contains these strings and the signal (string) J, the broadcast unit

$$*pI\nabla:J:pI_1\nabla$$

then accomplishes the task, with J signaling that the sample-taking procedure is to be initiated, and I_1 indicating the result. Simple, nonuniform random variables can be approximated by making multiple copies of numbers in the base (so that their proportions approximate the nonuniform distribution). More complex distributions can be handled by using the general computational powers of sets of broadcast units in conjunction with the above procedure.

3. The object is to generate a sample as in (2) but without replacement (the number drawn is deleted from the data base tagged by I). To accomplish this a second broadcast unit is added to the one in (2) giving

$$*pI\nabla:J:pI_1\nabla*pI_1\nabla::pI\nabla.$$

The second unit deletes the string just selected from the data base since just that string is uniquely prefixed with I_1 by the first broadcast unit.

4. A particular substring I_0 is to serve as a special punctuation mark and the object is to cleave an arbitrary string at the first (ith) occurrence of I_0 (if it occurs) in that string. To accomplish this let I be a prefix identifying the string to be cleaved, let I_1 identify the component to the left of I_0 after the cleavage and let I_2 identify the component to the right (including I_0). The following set of broadcast units accomplishes the cleavage at the first occurrence of I_0:

$*J_1:pI_1$
$*J_1:pI\nabla:pI_2\nabla$

Signal J_1 initiates the process and the string which will be developed into the right component is given its initial configuration, i.e., it is "set equal" to the string to be cleaved.

$*pI_2I_0\diamondsuit:J_2$
$*:pI_2I_0\diamondsuit:J_3$

A test is made to see if the punctuation I_0 is a prefix of the current version of the right component. If so signal J_2 is emitted, indicating that cleavage has been accomplished. Otherwise J_3 is emitted, indicating that the test should be repeated one place to the right.

$*J_1:J_4$
$*J_4:J_5$
$*J_5:J_3:J_6$

Signal J_6 indicates that the punctuation test failed exactly two time-steps ago.

$*pI_2\triangle\triangledown{:}J_6{:}I_3\triangledown$ $*I_3\triangle{:}pI_1\triangledown{:}pI_1\triangledown\triangle$	The left component is readied for a new test by having the leftmost symbol of the "old" right component added to its right end.
$*pI_2\triangle\triangledown{:}J_6{:}I_4\triangledown$ $*I_4\triangledown{:}pI_2\triangledown$	The right component is "updated" by deleting the symbol just added to the left component.
$*pI_2\triangledown{:}J_6{:}J_6\triangledown$ $*J_6\triangledown{:}{:}pI_2\triangledown$ $*pI_1\triangledown{:}J_6{:}J_7\triangledown$ $*J_7\triangledown{:}{:}pI_1\triangledown$	The "old" right and left components are deleted (simultaneously with the "updating" above) so that only the "new" components for testing appear in the state after two time-steps.
$*J_6{:}J_8$ $*J_8{:}J_4$	J_4 signals that the punctuation test is to be reinitiated for the "new" components.

To cleave the string at the *i*th occurrence of I_0, instead of the first, a count must be made of successive occurrences of I_0. Since the signal J_2 signals such an occurrence, this means counting successive occurrences of J_2, restarting the process each time the count is incremented (by issuing the signal J_3) until the count reaches i. The next example indicates how a binary counter can be set up to record the count.

5. The object is to count (modulo 2^n) occurrences of a signal J_2. The basic technique is illustrated by the construction of a one-stage binary counter. The transition function (table) for a one-stage binary counter is

Input at t	*State at t*	*State at t + 1*
J_2 not present	S_0	S_0
J_2 not present	S_1	S_1
J_2	S_0	S_1
J_2	S_1	S_0

The set of broadcast units which realizes this function for an initial state (signal) S_0 is:

$*{:}J_2{:}J_0$ $*J_2{:}J_1$	Makes current condition of input available for calculation of new state (on next time-step).
$*S_0{:}J_{10}$ $*S_1{:}J_{11}$	Makes current state available for calculation of new state (on next time-step).
$*J_0{:}J_{10}{:}S_0$ $*J_0{:}J_{11}{:}S_1$ $*J_1{:}J_{10}{:}S_1$ $*J_1{:}J_{11}{:}S_0$	Realization of the transition table.

For example, if J_2 occurs at times $t = 1$ and $t = 3$ the sequence of all signals broadcast (the overall state sequence) is:

t	*1*	*2*	*3*	*4*
Input signal	J_2		J_2	
Internal signals		J_1	J_0	J_1
	S_0	S_1	S_1	S_0
		J_{10}	J_{11}	J_{11}

The use of broadcast units to realize the given transition table is perfectly general and allows the realization of arbitrary transition functions (including counts modulo 2^n).

6. Treating the persistent strings as data implies that it should be possible to process them in standard computational ways. As a typical operation consider the addition of two persistent binary integers. The object, then, is to set up broadcast units which will carry out this addition. Let A_1 and A_2 be the suffixes which identify the two strings. The addition can be carried out serially, digit by digit, from right to left. Much as in example (4) the "rightmost" digits are successively extracted by the broadcast units

$$*\nabla\triangle A_1{:}I_1\triangle$$
$$*\nabla\triangle A_2{:}I_2\triangle$$

These digits, together with the "carry" from the operation on the previous pair of digits, identified by prefix I_3, are submitted as in example (5) to broadcast units realizing the transition table:

I_1 *Addend 1*	I_2 *Addend 2*	I_3 *Carry*	I_4 *Sum*	I_5 *New Carry*
0	0	0	0	0
0	0	1	1	0
0	1	0	1	0
0	1	1	0	1
1	0	0	1	0
1	0	1	0	1
1	1	0	0	1
1	1	1	1	1

Successive digits of the sum are assembled by the broadcast unit $*I_4\triangle{:}pA\nabla{:}pA\triangle\nabla$ where, at the end of the process, the prefix A designates the sum. A few additional "housekeeping" units such as $*pA\nabla{:}p\nabla A$, which puts the sum in the same form as the addends, are required to start up the process, keep track of position, etc. The overall process is simply a straightforward extension of techniques already illustrated.

7. As a final example note that any string identified with a suffix I can be *reproduced* by the broadcast unit $*\nabla I{:}\nabla I$. Note additionally that this unit itself has the suffix I! Hence, if we start with this unit alone, there will be 2^t copies of it after t time-steps. By revising the unit a bit, so that its action is conditional on a signal J, $*\nabla I{:}J{:}\nabla I$, this *self-reproduction* can be controlled from outside (say by other broadcast units). By extending this idea, with the help of the techniques outlined previously, we can put together a set of broadcast units which reproduces an arbitrary set of broadcast units (including itself). The result is a self-reproducing entity which can be given any of the powers expressible in the "broadcast language."

At this point it would not be difficult to give the "broadcast language" a precise, axiomatic formulation, developing the foregoing examples into a formal proof of its powers. (For anyone familiar with the material presented, say, in Arbib [1964] or Minsky [1967] this turns out to be little more than a somewhat tedious exercise.) However, our present objectives would be little advanced thereby. It is already reasonably clear that the "broadcast language" exhibits the desiderata outlined at the beginning of section 2. In particular, the broadcast units satisfy the functional integrity requirement (2) in a straightforward way. Consequently, strings of broadcast units can be manipulated by generalized genetic operators with attendant advantages vis-à-vis schemata (see section 6.3 and the close of chapter 7). Moreover a little thought shows that by using the techniques of usage (4) along with those of (2), units can be combined to define a crossover operator which acts only at specified "punctuations" (such as * s or : s or at a particular "indicator" string I). The other generalized genetic operators can be similarly defined. New detectors can be formed naturally from environmental signals (represented as binary strings). For example, a signal can be converted to an argument which will detect similar signals (elements of a superset) simply by inserting "don't cares" ($\Diamond$) at one or more points. Thus, $*E\nabla{:}pI\nabla$ converts any signal with prefix E ("environmental") into a permanent piece of data which can then be manipulated as in usages (4) and (6) to form a new broadcast unit with some modification of the signal as an *argument*.

The collection of broadcast units employed by an adaptive system at any time will, in effect, determine its representation of the environment. Since the units themselves are strings which can be manipulated by generalized genetic operators, strings of units ("devices") can be made subject to reproductive plans and intrinsic parallelism in processing. More than this, the adaptive plan can modify and coordinate the broadcast units to form models of the environment. By implementing, within the language, the "prediction and correction" techniques for models discussed in Illustration 3.4, we can arrive at a very sophisticated adaptive plan, one which can rapidly overcome inadequacies in its representation of the environment. This approach will be elaborated in the next section. The "broadcast language" already makes it clear that there exist languages suitable both for defining arbitrary representations and for defining the operators which allow these representations to be adapted to environmental requirements.

4. CONCERNING APPLICATIONS AND THE USE OF GENETIC PLANS TO MODIFY REPRESENTATIONS

The broadcast language provides unusually straightforward representations for a variety of natural models. Such representation not only provides a uniform context for comparisons and rigorous study, it also makes clear the "computational" or processing power of the model and its susceptibility to adaptation.

The Britten-Davidson generalization (1969) of the "operon-operator" model serves well to illustrate the point. This is a model for regulation in higher cells; as such it includes many mechanisms not found in the simpler bacterial cells modeled by Jacob and Monod (1961). The model consists of four basic types of gene (see also Figure 14):

1. The *sensor* gene is activated (perhaps via intermediary molecules) by any of various agents (enzymes, hormones, metabolites) involved in inter- and intracellular control. That is, the sensor gene is a detector sensitive to the state of the cell and its environment.
2. The *producer* genes are the specific controls for the actual production of cell structures (membranes, organelles, etc.) and operating agents (enzymes, etc.). They are the output controls of the regulation procedure.
3. Each *integrator* gene is associated with a sensor gene and sends out a specific signal (molecule) to other genes when the sensor is activated. Several integrator genes may be associated with a single sensor, thus allowing the sensor to initiate a variety of signals.
4. The *receptor* gene is a link between integrator genes and producer genes. Each receptor gene is associated with a single producer gene and is sensitive to a single integrator signal. When the signal is received the producer gene is activated. A given producer gene may have several associated receptors.

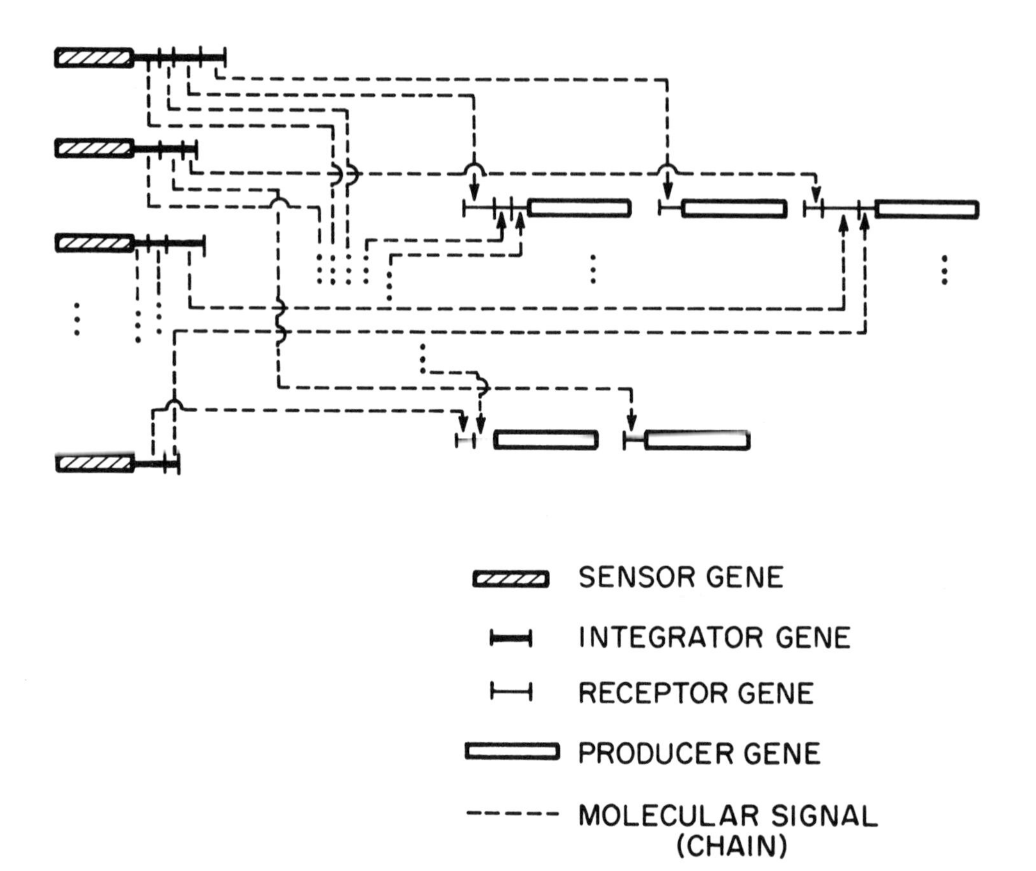

Fig. 14. Schematic of Britten-Davidson generalized "operon-operator" model for gene regulation in higher cells

Translated to the broadcast language a sensor-integrator gene complex $SI_1I_2I_3$ is directly represented by the set of broadcast units $*S{:}I_1*S{:}I_2*S{:}I_3$. A receptor-producer complex R_1R_2P is similarly represented by $*R_1{:}P*R_2{:}P$. If the receptor R_1 responds only to the integrator I_3, say, then $*R_1{:}P$ would be replaced by $*I_3{:}P$. It is apparent, however, that the receptor R_1 *could* be activated by several related signals, say *IA*, *IB*, *IC*, . . . , in which case the producer complex would be represented by $*I\diamondsuit{:}P$. Similarly a sensor S could be sensitive to any product with an initial radical X so that $SI_1I_2I_3$ could be represented by $*X\diamondsuit{:}I_1*X\diamondsuit{:}I_2*X\diamondsuit{:}I_3$. Clearly extremely complex feedback loops can be constructed, allowing a great range of conditional actions dependent upon substrate and products already produced. As an example (simplified for brevity) the sensor-integrator-receptor-producer complex $*X_1\diamondsuit:I_1A*I_1\diamondsuit:X_2P*X_2P:I_1C$ maintains production of X_2P if a metabolite with initial radical X_1 ever makes an appearance. In fact, if sensors and receptors can have the same range of sensitivity as the arguments of broadcast units, it is easy to show that there is an appropriate Britten-Davidson model for producing *any* arbitrarily given sequence of products.

A very similar representation can be produced for the lymphocyte immune network, well described by Niels K. Jerne (1973) and presented more technically by M. Sela (1973). In this case the "detectors" are combining sites on antibody molecules produced by lymphocyte cells. The environmental "signals" are invading antigens (e.g., foreign protein molecules). The presence of a detected antigen causes the production of additional lymphocytes (additional "broadcast units," see usage (7) of section 3) which in turn secrete additional antibodies which combine with (and neutralize) the antigens.

A bit further afield the broadcast language can also serve for a straightforward representation of the cell assembly model of the central nervous system (section 3.6). Here the broadcast units are cell assemblies while the "to-whom-it-may-concern" aspect of the broadcast language is reasonably approximated by the large number of neurons (10^3 to 10^4) in other assemblies contacted by *each* neuron in a given cell assembly. (More specific interconnections can be represented by appropriate "tagging" (prefixes) as in section 3.) Then, synaptic "learning" rules which induce fractionation and recruitment in cell assemblies find counterparts in generalized genetic operators which modify representations. Closely associated cell assemblies become the counterparts of tested representational components (cf. schemata), and so on. (The interested reader should consult Plum [1972] for the details of a related model.)

In the context of the broadcast language, the cell assembly model fits smoothly with the predictive modeling technique of section 3.4. A discussion of the latter implementation also gives an indication of how the broadcast language is applied to artificial systems. One implementation which emphasizes the cell-assembly/predictive-modeling fit relies on a set of *behavioral units* which generate action sequences and are modified on the basis of the outcome. Each behavioral unit consists of a population of *behavioral atoms* realized as devices in the broadcast language. If we look back to the search strategies of Figure 6 it is the detectors which have a role comparable to the atoms here. In the broadcast language, the detectors would be broadcast units (or sets of them) with arguments corresponding to the conditions defining the detector. (For example, the atom corresponding to δ_1 in Figure 6 would be activated by any 4-by-4 array with 8 or more dark squares.)

The unit(s) implementing the detector, when activated, would broadcast a signal with an identifying prefix. (For the reader familiar with the early history of pattern recognition these units act much like the demons at the lowest level of Selfridge's Pandemonium [1959].) Other devices would "weight" the signals, "sum" them, and "compare" the result to a threshold (cf. section 7.3) to determine which response signal (from the set of transformations $\{\eta_i\}$) should be broadcast. More generally, the behavioral atoms would be a string of broadcast units with an "initiate" condition C which specifies the set of signals capable of activating the atom, an "end" signal J which indicates the end of the atom's activity, and a "predicted value" signal which is meant to indicate the ultimate value to the behavioral unit of that atom's activation. With this arrangement we can treat the behavioral unit as a *population* of atoms. The atom activated at any given time is determined by a competition between whatever atom is already activated and all atoms having condition C satisfied by a signal in the current state $S(t)$. The higher the predicted value of the atom the more likely it is to win the competition. The object at this level is to have each atom's predicted value V consistent with that of its successor, so that a set of atoms acting in sequence provides a consistent prediction of their value to the behavioral unit. (In this way the atoms satisfy the error correction requirements discussed at the beginning of section 3.4 under element (iii) of a typical search plan.) This object can be accomplished via a genetic plan applied to the population of atoms—the reproduction of any atom is determined by the match between its predicted value and that of whatever atom is next activated. For example, consider two atoms, a_1 with parameters (C, J, V) and a_2 with parameters (C', J', V'), where a_1's end signal acts as a_2's initiate signal. Then $(V' - |V - V'|)$ could be used as a payoff to a_1 for purposes of the genetic plan, since the quantity measures the match between V' and V. The population would then be modified as outlined in section 6.1, new atoms being assigned the predicted value of the successor a_2. That is, the offspring of a_1 would be assigned the predicted value V'. All atoms active since the last actual payoff from the environment, and their offspring, are tagged and their predicted values are adjusted up or down at the time of the next environmental payoff. The adjustment is determined by the difference between predicted value and the actual payoff rate (payoff received from the environment divided by the elapsed time since last payoff). After each environmental payoff the active behavioral unit is subjected to a genetic plan (again as described in section 6.1). The behavioral unit next active (after the environmental payoff) is determined by the winner of a competition among *all* atoms in all behavioral units. The outcome of the competition is determined in the same way as the within-unit competition. Finally, a behavioral unit may be subjected to competition in the absence of environmental payoff, the probability of such an event increasing as a function of elapsed time since last payoff.

Much more detail would have to be supplied for this implementation of predictive modeling to reach the level of precision earlier given to the description of genetic plans. However the objective here is only to indicate the potential of the broadcast language for predictive modeling with changing representations.

As a final example, note that the world of radiative signals (sound, light, etc.) is susceptible to modeling as a complex broadcast system. In fact one physical realization of devices specified in the broadcast language would assign a unique frequency to each signal and realize broadcast units as a variety of frequency modulation devices.

Even where the broadcast language does not so directly represent extant models, it still supplies a rigorous framework for the description and modification of representations. In particular it makes possible the application of genetic plans to the problem of discovering suitable representations. Because devices are represented by strings and because the functional elements (the broadcast units) are self-defined, the generalized genetic operators of sections 6.2 and 6.3 can be used to modify the devices. Moreover, as indicated in section 3, these operators can themselves be defined within the broadcast language. This makes possible a hierarchy of operators defined with respect to various kinds of punctuation. Thus, one crossover operator could be defined to produce crossing-over anywhere along the string, another could be defined to produce crossing-over only at the symbol : (thereby providing for exchange of arguments between broadcast units), still another only at * (thereby exchanging broadcast units between devices), and so on. The operators so defined introduce a hierarchy of schemata ranging from schemata concerned primarily with varieties of arguments, through schemata concerned with combinations of broadcast units, and on to higher levels of organization (e.g., behavioral atoms, behavioral units, etc.).

Note that for the broadcast language schemata are generally defined with respect to sets of arbitrarily long strings. That is, the set of all devices specifiable in the broadcast language would be the set of *all* strings which can be formed from the ten basic symbols; since a schema designates the set of all strings which match it on its defining positions, each schema designates a countable subset of devices. Using this extension of the notion of a schema, we see that the results of chapters 6 and 7, particularly those pertaining to intrinsic parallelism, extend directly to the adaptation of codings and representations. Since the operators themselves can also be specified within the broadcast language, they can also be made subject to the same adaptive processes.

In sum: This chapter has been concerned with removing the limitations imposed by fixed representations. To this end it is possible to devise languages—the broadcast language is an example—which use strings to rigorously define all effectively specifiable representations, models, operators, etc. Since the objects of the language are presented as strings, they can be made grist for the mill provided by genetic plans. As a consequence the advantages of compact storage of accrued information, operational simplicity, intrinsic parallelism, robustness, etc., discussed in chapters 6 and 7, extend to adaptation of representations.

9. An Overview

Enough of the theoretical framework has now been erected that we can begin to view it as a whole. To this end, the present chapter will discuss three general aspects of the theory. Section 1 will concentrate on those insights offered by the theory which are useful across the full spectrum of adaptive problems. Section 2 provides a synopsis of several computer studies to give the reader an idea of how the overall theory works in particular contexts. Section 3 will outline several difficult long-range problems which fall within the scope of the theory.

1. INSIGHTS

Within the theoretical framework problems of adaptation have been phrased in terms of generating structures of progressively higher performance. Because the framework itself places no constraints on what objects can be taken as structures, other than that it be possible to rank them according to some measure of performance, the resulting theory has considerable latitude. Once adaptation has been characterized along these lines, it is also relatively easy to describe several pervasive, interrelated obstacles to adaptation—obstacles which occur in some combination in all but the most trivial problems:

1. High cardinality of $\mathcal{A}$. The set of potentially interesting structures is extensive, making searches long and storage of relevant data difficult.
2. Apportionment of credit. Knowledge of properties held in common by structures of above-average performance is incomplete, making it difficult to infer from past tests what untested structures are likely to yield above-average performance.
3. High dimensionality of μ_E. Performance is a function of large numbers of variables, making it difficult to use classical optimization methods employing gradients, etc.

4. Nonlinearity of μ_E. The performance measure is nonlinear, exhibiting "false peaks" and making it difficult to avoid concentration of trials in suboptimal regions.
5. Mutual interference of search and exploitation. Exploitation of what is known (generation of structures observed to give above-average performance) interferes with acquisition of new information (generation of new structures) and vice versa.
6. Relevant non-payoff information. The environment provides much information in addition to performance values (payoff), some of which is relevant to improved performance.

The schema concept suggests a coordinated array of robust procedures for meeting these obstacles. The procedures are all founded on the view that each structure is a "carrier" (or selected sample point) of each of the great number of schemata it instances. Because arbitrary structures are easily represented as strings (by using detectors or more sophisticated techniques such as the broadcast language) the resulting procedures apply to adaptation in all its forms. Once schemata have been defined, there is a natural means (p. 69) of comparing structures and apportioning credit by assigning to each schema the average of payoffs to its *observed* instances (compensating obstacle (2)). A small population of structures, when properly selected (pp. 139–40), can then store the relative performance rankings for very large numbers of schemata (compensating obstacle (1)). It is this broad data base vis-à-vis schemata (p. 87) which enables genetic plans to escape false peaks and other difficulties engendered by nonlinearities (compensating obstacle (4)). Recasting the search problem in terms of the space of schemata sidesteps dimensionality effects (obstacle (3)), at least for intrinsically parallel procedures such as genetic plans (p. 71). Under such plans the succession of structures generated from the data base (the current population) induces a highly parallel, diffusion-like spread of trials in the space of schemata (pp. 104–6). This takes place in such a fashion that there is:

(1) progressive exploitation of the best observed schemata,
(2) increasing confidence in the estimates of the expected payoff to the best observed schemata,

and (3) testing of great numbers of new combinations of schemata (both newly generated schemata and new combinations of already tested schemata of high rank).

In the particular case of the interaction of crossover and inversion with reproduction a net of associations is induced (p. 108). Coadapted attributes (attributes defining schemata of above-average performance) become tightly linked and increase their proportion in the population (p. 127). In fact (p. 137) the expected rate of increase dP_ξ/dt of the proportion of any given schema ξ is closely approximated by

$$dP_\xi/dt = P_\xi(\mu_\xi - \bar{\mu}) = P_\xi \alpha_\xi,$$

where α_ξ is the average excess of the random variable ξ in the population $\mathcal{B}(t)$. This formula is analogous to Fisher's (1930) classical result for single alleles and reduces to it when ξ is restricted to a single defining position. The resulting intrinsic parallelism greatly ameliorates the conflict between search and exploitation (obstacle (5)). By building up representations and models in terms of a language like the broadcast language (p. 152) the overall advantages of the schema approach can also be brought to bear on the problem of non-payoff information (obstacle (6)). The schemata provide for apportionment of credit to various aspects of the model on the basis of their relevance to realized predictions.

2. COMPUTER STUDIES

At the time of this writing several computer studies of genetic plans have been completed (and more are underway). Four studies closely related to the theoretical framework will be outlined here: R. S. Rosenberg's *Simulation of Genetic Populations with Biochemical Properties* (1967), D. J. Cavicchio's *Adaptive Search Using Simulated Evolution* (1970), R. B. Hollstien's *Artificial Genetic Adaptation in Computer Control Systems* (1971), and D. R. Frantz's *Non-linearities in Genetic Adaptive Search* (1972).

Richard Rosenberg completed his computer study of closed, small populations while formulation of the theoretical framework was still in its early stages. He concentrated on the complicated relationship between genotype and phenotype under dynamic interaction between the population and its environment. The model's central feature is the definition of phenotype by chemical concentrations. These concentrations are controlled by enzymes under genetic control. Epistasis has a critical role because several enzymes (and hence the corresponding genes) can affect any given phenotypic characteristic (chemical concentration). Though the variety of molecules, enzyme-controlled reactions, and genes is kept small to make the study feasible, it still presents a detailed picture of the propagation of advantageous, linked genes through a small population. Moreover the study suggests general relations between the number of genes, crossover probabilities, and the rate of adaptation under epistasis. Equally important, the study makes it clear that quite complex ("molecular") definitions of phenotype can be simulated without losing relevance, up to and including suggestions for experiments *in vivo* and *in vitro*. (At least two subsequent detailed studies of biological cells were directly encouraged by this experience, R. Weinberg's *Computer Simulation of a Living Cell* [1970] and E. D. Goodman's *Adaptive Behavior of Simulated Bacterial Cells Subjected to Nutritional Shifts* [1972].)

The first study based directly upon the theoretical framework was that of Daniel Cavicchio. (J. D. Bagley's *The Behavior of Adaptive Systems Which Employ Genetic and Correlation Algorithms* [1967] is an earlier study which is a direct precursor of both this study and Frantz's.) The set of structures $\mathcal{A}$ is taken to be a broad class of pattern classification devices based on those developed by Bledsoe and Browning (1959) and Uhr (1973). Specifically each device uses a set of detectors to process information presented by the sensors in a 25 by 25 array (cf. section 1.3 and Figures 5 through 7). After an initial "training" period, during which the device $A \in \mathcal{A}$ accumulates information about one or more handwritten alphabets, A is tested and scored on its classification of letters from another handwritten alphabet. This score amounts to A's performance rating, its payoff $\mu(A)$. The adaptive plan, a version of the $\mathcal{R}_d$ class of reproductive plans (pp. 94–95), generates new detectors (and, in the process, new devices) by using genetic operators which are variations on the operators discussed in sections 6.2 through 6.4.

Because of the sophistication of the problem environment, the first objective is to develop some estimate of the task's difficulty vis-à-vis the devices in $\mathcal{A}$. Cavicchio does this by testing, in the problem setting, a large number of devices drawn at random from $\mathcal{A}$. The observed distribution of performances is Gaussian. For a typical environment (Cavicchio calls it the "difficult task"), the mean score is 17 with a standard deviation of 5. (A perfect score would be 100.) This implies that in 1000 random trials of devices drawn from $\mathcal{A}$ we can expect the best performance to be about 32.

To obtain an idea of the performance of a nonreproductive, but adaptive, plan in the same environment, Cavicchio applied a version of Uhr and Vossler's (1973) "detector evaluation" procedure to the search of $\mathcal{A}$. This procedure amounts to identifying inferior detectors and replacing them with "mutated" versions. The *best* performance observed over a great many runs of 600 trials each was a score of 52; each of the runs "leveled out" somewhere between the 300th and the 600th trial. This is considerably better than a random search, being 7 standard deviations above the mean in less than 600 trials (as compared to 3 standard deviations in 1000 trials).

Against this background Cavicchio then developed and tested a series of reproductive plans. The best of these attained a score of 75.5 in 780 trials, a score considerably beyond that attained in any of the "detector evaluation" runs. (In qualitative terms, a score of 52 would correspond to a "poor" human performance, while a score of 75.5 would correspond to a "good" human performance. Because many characters in the "difficult task" are quite similar in form, increments in scoring are difficult to attain after the easily distinguished characters have been handled.) An important general observation of this study is that the sophistication and power of a genetic plan is lost whenever M, the size of the population (data base), is very small. It is an overwhelming handicap to use only the most recent trial ($M = 1$) as the basis for generating new trials (cf. Fogel et al. 1966). On the other hand, the population need not be large to give considerable scope to genetic plans (20 was a population size commonly used by Cavicchio).

Roy Hollstien added considerably to our detailed understanding of genetic plans by making an extensive study of genetic plans as adaptive control procedures. His emphasis is on domains wherein classical "linear" and "quadratic" approaches are unavailing, i.e., domains where the performance function exhibits discontinuities, multiple peaks, plateaus, elongated ridges, etc. To give the problems a uniform setting he transforms them to discrete function optimization problems, encoding points in the domain as strings (see p. 57). An unusual and productive aspect of Hollstien's study is his translation of breeding plans for artificial genetic selection into control policies. A breeding plan which employs inbreeding within related (akin) policies, and recurrent crossbreeding of the best policies from the best families, is found to exhibit very robust performance over a range of 14 carefully selected, difficult test functions. (The test functions include such "standards" as Rosenbrock's ridge, the sum of three Gaussian 2-dimensional density functions, and a highly discontinuous "checkerboard" pattern.) The test functions are represented on a grid of 10,000 points (100 by 100). In each case the region in which the test function exceeds 90 percent of its maximum value is small. For example, test function 7 with *two* false peaks (the sum of three Gaussian 2-dimensional densities) exceeds 90 percent of its maximum value on only 42 points out of the 10,000. The breeding plans are tested over 20 generations of 16 individuals each, special provisions being made to control random effects of small sample size ("genetic drift"). The breeding plan referred to above, when confronted with test function 7, placed *all* of its trials in the "90 percent region" after 12 generations (192 trials). A random search would be expected to take 250 trials

(10,000/42) to place a *single* point in the "90 percent region." The same breeding plan performs as well or better on the 13 other test functions. Given the variety of the test functions, the simplicity of the basic algorithms, and the restricted data base, this is a striking performance.

Daniel Frantz concentrated on the internal workings of genetic plans, observing the effect, upon the population, of dependencies in the performance function. Specifically, he studies situations in which the quantity to be optimized is a function of 25 binary parameters. I.e., $\mathcal{E}$ consists of functions which are 25-dimensional and have a domain of $2^{25} = 3.2 \times 10^7$ points. Dependencies between the parameters (nonlinearities) are introduced to make it impossible to optimize the functions dimension by dimension (unimodality is avoided). Frantz's procedure is to detect the effects of these dependencies upon population structure (gene associations) by using a multidimensional chi-square contingency table. As expected from theoretical considerations (see Lemma 7.2 and the discussion following it) algebraic dependencies (between the parameters) induce statistical dependencies (between alleles). That is, in the population, combinations of alleles associated with dependent parameters occur with probabilities different from the product of the probabilities of the individual alleles. Moreover there is a positional effect on the rate of improvement: For functions with dependencies the rate of improvement is significantly greater when the corresponding alleles are close together in the representation. This effect corresponds to the theoretical result that the ability to pass good combinations on to descendants depends upon the combinations' immunity to disruption by crossover. It is significant that, for the problems studied, the optimum was attained in too short a time for the inversion operator to effectively augment the rate of improvement (by varying positional effects).

3. ADVANCED QUESTIONS

The results presented in this book have a bearing on several problem areas substantially more difficult than those recounted in section 9.1. Each of these problems has a long history and is complex enough to make sudden resolution unlikely. Nevertheless the general framework does help to focus several disparate results, providing suggestions for further progress.

As a first example, let us look at the complex of problems concerned with the dynamics of speciation. These problems have their origin in biology, but a close look shows them to be closely related to problems in the optimal allocation of limited resources. To see this, consider the following idealized situation. There are two one-armed bandits, bandit ξ_1 paying 1 unit with probability p_1 on each trial, bandit ξ_2 paying 1 unit with probability $p_2 < p_1$. There are also M players. The casino is so organized that the bandits are continuously (and simultaneously) operated, so that at any time t, for a modest fee, a player may elect to receive the payoff (possibly zero) of one of the two bandits. The manager has, however, introduced a gimmick. If M_1 players elect to play bandit ξ_1 at time t, they must *share* the unit of payoff if the outcome is successful. That is, on that particular trial, each of the M_1 players will receive a payoff of $1/M_1$ with probability p_1. Now, let us assume that the M players *must* participate for a period of T consecutive trials. If there is but one player ($M = 1$), clearly he will maximize his income (or minimize his losses) by playing bandit ξ_1 at all times. However, if there are $M > 1$ players the situation changes. There will be stable queues, where no player can improve his payoff by shifting from one bandit to another. These occur when the players distribute themselves in the ratio $M_1/M_2 = p_1/p_2$ (at least as closely as allowed by the requirement that M_1 and M_2 be integers summing to M). For example, if $p_1 = \frac{1}{2}$, $p_2 = \frac{1}{8}$, and $M = 10$, there will be 8 players queued in front of bandit ξ_1 and 2 players in front of bandit ξ_2. We see that with limited resources (in the numerical example, a maximum of 2 units payoff per trial and an expectation of $\frac{5}{8}$ unit) the population of players must divide into two subpopulations in order to optimize individual shares of the resources (the "bandit ξ_1 players" and the "bandit ξ_2 players"). Similar considerations apply when there are $r > 2$ bandits.

We have here a rough analogy to the differentiation of individuals (the subpopulations) to exploit environmental niches (the bandits). The analogy can be made more precise by recasting it in terms of schemata. Let us consider a population of M individuals and the set of 2^{l^0} schemata defined on a given set of l^0 positions. Assume that schema ξ_i, $i = 1, \ldots, 2^{l^0}$, exploits a unique "environmental niche" which produces a total of Q_i units of payoff per time-step. (Q_i corresponds to the renewal rate of a critical, volatile resource exploited by ξ_i.) If the population contains M_i instances of ξ_i, the Q_i units are shared among them so that each instance of ξ_i receives a payoff of Q_i/M_i. Let $Q_{(1)} > Q_{(2)} > \ldots > Q_{(2^{l^0})}$ so that schema $\xi_{(1)}$ is associated with the most productive niche, $\xi_{(2)}$ with the second most productive niche, etc. Clearly when $M_{(1)}$ is large enough that $Q_{(1)}/M_{(1)} < Q_{(2)}$, an instance of $\xi_{(2)}$ will be at a reproductive advantage. Following the same line of argument as in the case of the 2 one-armed bandits, we get as a stable distribution the obvious generalization:

$$M_{(i)} = cQ_{(i)}/Q_{(j)}$$

where j is the smallest index such that

$$\sum_{i=1}^{j+1} Q_{(i)}/Q_{(j+1)} > M$$

and c is chosen so that

$$\sum_{i=1}^{j} cQ_{(i)}/Q_{(j)} = M$$

(modified so that the actual solution is in integers). For example, let $l^0 = 2$ with 2 alleles (attributes) at each locus, yielding schemata ξ_1, ξ_2, ξ_3, ξ_4 with $Q_1 = 1$, $Q_2 = 4$, $Q_3 = 8$, $Q_4 = 1$. Then for $M = 9$ there will be 6 instances of ξ_3, 3 instances of ξ_2, and no instances of ξ_1 or ξ_4 in the stable distribution.

Here we have a simple example of speciation. If the population is restricted to M individuals (by factors other than the niche payoff rates), certain combinations of alleles appear in a stable competition while other combinations are proscribed by the same competition. The example can rapidly be made more realistic by letting the payoff to each schema ξ be a random variable with *expected* payoff

$$\mu_\xi(t) = \min \{\mu_\xi^0, Q_\xi/M_\xi(t), Q/M(t)\}$$

where Q_ξ is the minimum of the renewal rates of resources characterizing the environmental niche associated with ξ, $M_\xi(t)$ is the number of instances of ξ at time t, Q is the minimum of the renewal rates of resources required by all the schemata, and $M(t)$ is the total population at time t. Now the schema ξ will increase its proportion at an intrinsic rate set by μ_ξ^0 *until* it reaches the "carrying capacity" of its niche, determined by Q_ξ, or until the total population has increased to a point that the overall "carrying capacity," determined by Q, limits further expansion. (For the reader familiar with MacArthur and Wilson's [1967] work, the effect of Q_ξ corresponds to a K selection—crowded niche—effect, whereas μ_ξ^0 is the intrinsic rate of increase, possibly wasteful of resources, under classical r selection. Q sets an ultimate limit on the carrying capacity of the environment, no matter what the diversity or organization of the species.) With typical values for the $\{Q_\xi\}$ and Q, the population will once again develop into subpopulations characterized by certain combinations of alleles (schemata), with many combinations being proscribed.

The really interesting form of this theory would characterize niches (and hence the overall payoff function μ) in terms of the varieties of schemata that could exploit them—different schemata exploiting a given niche with differing efficiencies. The dynamics of speciation would then be determined by competition within and across niches. It is interesting that under these circumstances speciation *could* take place in the absence of isolation (in contrast to the usual view, cf. Mayr 1963).

Once an adaptive system discovers that given combinations of genes (or their alleles) offer a persistent advantage, new modes of advance become possible. If the given combinations can be handled as units they can serve as components ("super genes") for higher order units. In effect the system can ignore the *details* underlying the advantage conferred by a combination, and operate simply in terms of the advantage conferred. By so doing the system can explore regions of $\mathcal{A}$, i.e., combinations of the new units, which would otherwise be tried with a much lower probability. (For example, consider two combinations of 10 alleles each under the steady state of section 7.2. If *each* of the alleles involved occurs with a frequency of 0.8, the overall combination of 20 alleles will occur with a frequency $(0.8)^{20} \cong 0.01$. On the other hand, if each of the two 10 allele combinations is maintained at a frequency of only 0.5, then the 20-allele combination will occur with frequency $(0.5)^2 = 0.25$. I.e., the expected time to occurrence will be reduced by a factor of 25.) Since combinations of advantageous units often offer an advantage beyond that of the individual units—as when the units' effects are additive (linear independence) or cooperative—they are good candidates for early testing. (The cooperative case where one unit effects an enrichment which can be exploited by another is particularly common; cf. cooperating cell assemblies or stages of a complex production activity such as illustrated in Figure 3.)

We have already discussed (section 6.3) the way in which inversion can favor association between genes. However, by controlling representation, the adaptive system can bring about changes which go much further, producing a hierarchy of units. The basic mechanism stems from the introduction of arbitrary punctuation marks to control operators (see usage (4) in section 8.3 and the discussion on pages 152–53). The adaptive system introduces a distinct punctuation mark (specific symbol string) to mark off the combinations which are to be treated as units at a given level of the hierarchy. Then the operators for that level are restricted to act only at that punctuation. (E.g., crossover takes place only at the positions marked by the given punctuation.) By introducing another punctuation mark to treat *combinations* of these units, in turn, as *new* units, and so on, the hierarchy can be extended to any number of levels. The resulting structure offers the possibility of quickly pinpointing responsibility for good or bad performance. (E.g., a hierarchy of 5 levels in which each unit is composed of 10 lower level units allows any one of 10^5 components to be selected by a sequence of 5 tests.) In the hierarchy, the units at each level are subject to the same "stability" considerations as schemata (pp. 100–102), being continually modified by operators at lower levels. Thus certain hierarchies will be favored because of their stability, the corresponding punctuations and operators becoming common features of the overall population. Chapter 4 of Simon's book, *The Sciences of the Artificial* (1969), gives a good qualitative discussion of this and related topics.

It is natural to ask whether these operator-induced hierarchies can account for important features of such observed hierarchies as the organelle, cell, organ, organism, species, . . . hierarchy of biology, or the hierarchical organization of the CNS or a computer program. There would seem to be a strong relation between operator-induced hierarchies and the sequences of developmental biology (embryogenesis and morphogenesis) whereby, for example, a fertilized egg develops into a mature multicellular organism.

As a final problem area we can look to situations wherein payoff to a given structure varies in time and space. For example, in the case of limited resources, the resource renewal rates Q_ξ may be both temporally and spatially inhomogeneous, being described by a function $Q_\xi(x_1, \ldots, x_k, t)$. In such cases we would also expect the population at time t to be distributed spatially, yielding $\mathcal{B}(x_1, \ldots, x_k, t)$ as the component at coordinate $(x_1, \ldots, x_k)$. After some adaptation any one component of the population, in response to the spatial variations in payoff, will generally exhibit different proportions of schemata than its neighbors.

In ecological situations, as well as in certain control situations, it is appropriate to consider the migration of structures from one component of the population to another (one coordinate to another). That is, under the direction of the adaptive plan, the jth structure $A_j(x_1, \ldots, x_k, t)$ in the population component $\mathcal{B}(x_1, \ldots, x_k, t)$ may be transferred to a neighboring coordinate $(x_1', \ldots, x_k')$, becoming an element of $\mathcal{B}(x_1', \ldots, x_k', t+1)$. (Such systems can be usefully described with the help of cellular automata; see R. F. Brender's *A Programming System for Cellular Spaces* 1969 and *Essays on Cellular Automata* edited by A. W. Burks 1970.) Under these conditions we would expect to observe a *spatial* diffusion of schemata. Thus schemata having a large number of instances in $\mathcal{B}(x_1, \ldots, x_k, t)$ would be expected to appear in fair numbers in neighboring components of the population, even if their performance *there* is poor. At the "boundaries" between different niches the genetic operators will produce unusual "hybrids" of schemata common in each of the niches. That is, where there are sharp changes in the $Q_\xi(x_1, \ldots, x_k, t)$, crossover will yield a wide range of new schemata, which would otherwise occur with low probability. Many of these schemata will be unfit or fit only in the boundary region, but some may exhibit exceptional performance on one or both niches. The relation to Mayr's (1963) description of speciation as the result of contact between previously isolated, locally adapted populations is manifest. (See, however, the comment on page 166.) There is much to be learned about these processes, particularly with reference to schemata or coadapted sets. (Some of the most interesting work to date has been carried out by A. Brues 1972.) It is clear that the addition of migration rules to reproductive plans affords a sophisticated approach to spatially inhomogeneous environments, but we need to know a great deal more about the efficiency and robustness of such an approach (paralleling the development of chapters 5 and 7 for the homogeneous case).

So far we have been discussing spatial inhomogeneity of payoff, but temporal inhomogeneity or *nonstationarity* is an even more difficult problem. There are four points at which the results of this book have a bearing on such problems. First, and most obvious, the rapid response of reproductive plans, exhibited concretely in the studies of Cavicchio (1970) and Hollstien (1971), permits "tracking" of the changing payoff function. As long as the payoff function changes at a rate comparable to the response rate, overall performance will be good. The proportions of schemata in the population will change rapidly enough to take advantage of current features of the environment. As a second point, it should be noted that the rank bestowed on a schema (its proportion in successive generations) is the geometric mean of the observed averages $\hat{\mu}_\xi(t)$ (see Lemma 7.2). Thus more rapid fluctuations will favor schemata which exhibit the best (geometric) mean performance when subjected to the fluctuations. Third, if there are repetitive (not necessarily cyclic) features over time, dominance change provides a mechanism for retaining useful schemata when the features are not in force (see pages 115–16). By occasionally (say once every few generations) giving recessive status to *instances* of currently favored schemata, they can be reserved against adverse environmental configurations. In particular, these recessive instances have a much reduced testing rate (see page 115). As a result the recessive versions are relatively unaffected by environmental changes which quickly eliminate the dominant version. By occasionally returning an *instance* of a recessive schema to dominant status it can be tested against the current environmental configuration. If the dominant instance achieves above-average performance it will reproduce rapidly, producing an increasing proportion of dominant instances in the population. (If the performance is below average the newly dominant instance will quickly disappear, at no great cost to the efficiency of the adaptive plan.) Finally, by making the intrachromosomal duplication of a schema ξ subject to the disappearance of an environmental feature currently exploited by ξ, the effective mutation rate of ξ can be increased. For example, let the schema ξ be associated with a *sensor* (see pages 153–54) which detects the environmental feature exploited by ξ. Let intrachromosomal duplication be an operator controlled by the sensor; i.e., whenever the sensor is deactivated, intrachromosomal duplication takes place on the sets of genes associated with the sensor. In consequence, disappearance of the environmental feature will result in many copies of the genes, and hence the schemata, associated with the sensor. With a fixed mutation rate for each gene, the number of mutants of a given schema in the population will depend upon the number of copies thereof. Thus by providing many copies within a chromosome, the effective mutation rate is correspondingly increased. As a result, this (hypothetical) mechanism provides many variants relevant to the crisis. At the same time it retains whatever advantage remains to the original schema ξ. In biology there are varying amounts of evidence for the foregoing responses to nonstationarity, and some of the predicted effects have been demonstrated in simulations, but again we are a long way from a theory, or even good experimental confirmation, of their efficiency.

In these nine chapters we have come only a short way in the study of adaptation as a general process. The book's main objective has been to make it plausible that simple mechanisms can generate complex adaptations; however, the book will have fulfilled its role if it has communicated enough of adaptation's inherent fascination to make the reader's effort worthwhile.

10. Interim and Prospectus

Adaptation in Natural and Artificial Systems, after a seven-year gestation, made its appearance in 1975. It is now 1991, and much has happened in the interim. Topics that were speculative in 1975 have been carefully explored; extensions, applications, and new areas of investigation abound. More than 150 papers were submitted to the 1991 International Conference on Genetic Algorithms (Belew and Booker 1991), and several new books have been written about genetic algorithms (e.g., Davis 1987 or Davis 1991). There is even a textbook (Goldberg 1989). Most of this new research has been reported in the published proceedings of the genetic algorithm conferences of 1985, 1987, 1989, and 1991 (Grefenstette 1985, Grefenstette 1987, Schaffer 1989, and Belew and Booker 1991) and is readily accessible there, so I will not attempt to review it here—the review would be, at best, little more than an annotated listing. Instead, I'll follow the pattern of the rest of the book, using this new chapter to report on lines of research I've pursued since 1975. A new edition also provides an opportunity to correct errors in the original edition. Most of these are simple and innocuous, but an error in one proof, discovered and corrected by Dr. Daniel Frantz, is subtle and important. By good fortune, after the correction the theorem involved stands as stated. Finally, a new chapter offers an opportunity to look further into the future; this too I'll attempt.

1. IN THE INTERIM

Classifier Systems

Classifier systems, a specialization of chapter 8's "broadcast language," are a vehicle for using genetic algorithms in studies of machine learning. Classifier systems were introduced in Holland 1976 and were later revised to the current "standard" form in Holland 1980. There is a comprehensive description of the standard form, with examples, in Holland 1986, but there are now many variants (see Belew and Booker 1991). A classifier system is more restricted than the broadcast language in just one major respect: A broadcast unit can directly create other broadcast units, but a classifier, the broadcast unit's counterpart in a classifier system, cannot directly create other classifiers. This restriction permits a much simpler syntax based on only three atomic symbols, {1,0, # ("don't care")}. A classifier system creates new classifiers through the action of the genetic algorithm on the system's population of classifiers.

Classifier systems aim at a question that seems to me central to a deeper understanding of learning: How does a system improve its performance in a perpetually novel environment where overt ratings of performance are only rarely available? A learning task of this kind is more easily described if we think of the system as playing a strategic game, like checkers or chess. After a long sequence of actions (moves), the system receives some notification of a "win" or a "loss" and, perhaps, some indication of the strength of the win or loss. But there is almost no information about what moves should have been changed to yield better performance. Most learning situations for animals, including humans, have this characteristic—an extended sequence of actions is followed by some general indication of the level of performance, with little information about specific changes that would improve performance.

In defining classifier systems (see Figure 15), I adopted the common view that the state of the environment is conveyed to the system via a set of *detectors* (e.g., rods and cones in a retina). The outputs of the detectors are treated as standardized packets of information—*messages*. Messages are used for internal processing as well, and some messages, by directing the system's *effectors* (e.g., its muscles), determine the system's actions upon its environment. Beside the interactions with the environment provided by detectors and effectors, there is a further interaction that is critical to the learning process. The environment must, upon occasion, provide the system with some measure of its performance. Here, as earlier, I will use the term *payoff* as the general term for this measure.

The computational basis for classifier systems is provided by a set of *condition/action* rules, called *classifiers*. To simplify the computational basis, all interactions between rules are mediated by messages. Under this provision a typical rule, under interpretation, would have the form

> IF there is (a message from the detectors indicating) an object left of center in the field of vision
> THEN (by issuing a message to the effectors) cause the eyes to look left.

That is, the *condition* part of the rule "looks for" certain kinds of messages, and when the rule's conditions are satisfied, the *action* part specifies a message to be sent. Messages both pass information from the environment and provide communication

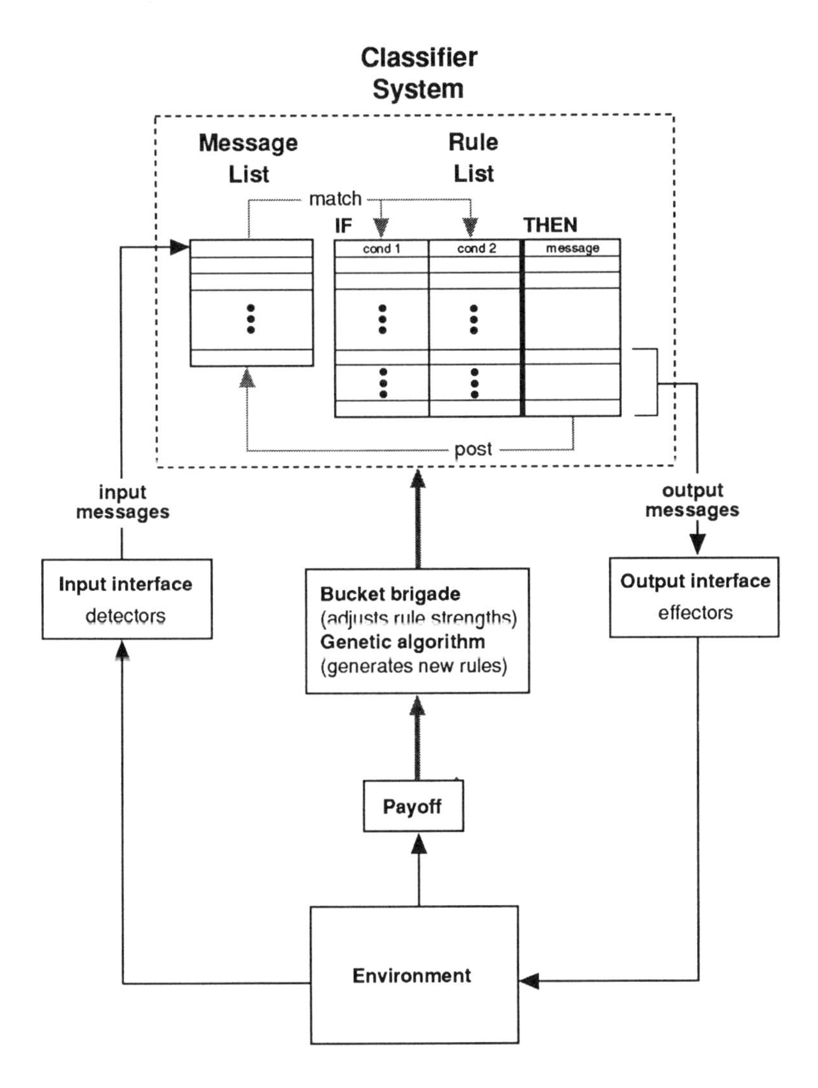

Fig. 15. A classifier system

between rules, as in the broadcast language (where they were called *signals*). Thus each rule is a simple message processor. Many rules can be active simultaneously, so many messages may be present at any given instant. It is convenient to think of the messages as collected in a *list* that changes under the combined impetus of the environment and the rules.

Parallelism, the concurrent activity of many rules, is an important aspect of classifier systems. Parallelism makes it possible for the system to combine rules into clusters that model the environment, providing two important advantages:

1. Combinatorics work for the system instead of against it. The system builds a "picture" of the situation from parts, rather than treating it as a monolithic whole. The advantage is similar to that obtained when one describes a face in terms of component parts. Select, say, 8 components for the face—hair, forehead, eyebrows, eyes, cheekbones, nose, mouth, and chin. Allow 10 variants for each component part—different hair colors and textures, different forehead shapes, and so on. Then $10^8 = 100$ million faces can be described by combining these components in different ways. This at a cost of storing only $8 \times 10 = 80$ "building block" components.
2. Experience can be transferred to novel situations. On encountering a novel situation, such as a "red car by the side of the road with a flat tire," the system activates several relevant rules, such as those for "red," "car," "flat tire," etc. When "building-block" rules, such as those for "car," have proved useful in past combinations, it is at least plausible that they will prove useful in new, similar combinations. To exploit these possibilities, the rules must be organized in a way that permits clusters of rules to be activated in changing combinations, as dictated by changing situations. Building-block rules then give the system a capacity for transferring experience to new situations.

To define a standard classifier system, we first require all messages to be bit-strings of the same length, k, much as one sets the register size for a computer. Formally, then, messages belong to the set $\{1,0\}^k$. The condition part of a rule is specified by the use of a "don't care" symbol, #, reminiscent of the "don't care" used to define schemata. Thus, the set of all conditions is the set $\{1,0,\#\}^k$. For k=6, the condition 1##### is satisfied by any message that starts with a 1, while the condition 001001 is satisfied by one and only one message, the message 001001. It is worth noting that a condition's *specificity* (the reciprocal of the number of messages that satisfy it) depends directly upon the number of #s in the condition—the more #s, the lower the specificity.

In the standard system, all rules consist of two conditions and a single outgoing message, which is sent when the two conditions are satisfied. Rules are specified in the format

1#####,001001/000011.

This format is interpreted as follows:

> IF condition 1 is satisfied (in this case, by a message, on the message list, that starts with 1),
> AND condition 2 is satisfied (in this case, by a second, specific, message 001001),
> THEN the message in the action part (in this case, 000011) is posted to the message list on the next time-step.

Conditions may be negated: For example, –1##### is satisfied if there is *no* message on the message list that begins with a 1. With these provisions it is easy to show that a classifier system is computationally complete, in the sense that any program that can be written in a standard programming language, such as Fortran, C, or Lisp, can be implemented within a classifier system.

Without any changes to this definition, rules can be given an "address" that can be used by other rules when that is useful. Consider a rule *r* with a condition of the form 111# . . . #. Any message that starts with three 1s will satisfy this condition. If this particular prefix, 111, is reserved for the rule *r* alone, then any message with that prefix will be directed to *r* and only to *r*. Such reserved prefixes (they can also be suffixes, or indeed any part of the message) are called *tags*. Of course, several rules might have the same reserved tag; that simply means that all of them receive messages so tagged, acting as a cluster with respect to that tag. Appropriate use of tags also permits rules to be coupled to act sequentially.

The basic execution cycle of the classifier system consists of an iteration of the following steps:

1. Messages from the environment are placed on the message list.
2. Each condition of each classifier is checked against the message list to see if it is satisfied by (at least one) message thereon.
3. All classifiers that have both conditions satisfied participate in a *competition* (to be discussed in a moment), and those that win post their messages to the message list.
4. All messages directed to effectors are executed (causing actions in the environment).
5. All messages on the message list from the *previous* cycle are erased (i.e., messages persist for only a single cycle, unless they are repeatedly posted).

Because the message list can hold an arbitrary number of messages, any number of rules can be active simultaneously; because the messages are simply uninterpreted bit-strings, there are no *consistency* problems in the internal processing. Consistency problems do arise at the effectors; when different, simultaneous messages urge an effector to take mutually exclusive actions, they are resolved by competition.

Competition plays a central role in determining just which rules are active at any given time. To provide a computational basis for the competition, each rule is assigned a quantity, called its *strength*, that summarizes its average past usefulness to the system. We will see shortly that the strength is automatically adjusted by a *credit assignment* algorithm, as part of the learning process. Competition allows rules to be treated as hypotheses, more or less confirmed, rather than as incontrovertible facts. The strength of a rule corresponds to its level of confirmation; stronger rules are more likely to win the competition when their conditions are satisfied. Stated another way, the classifier system's reliance upon a rule is based upon the rule's average usefulness in the contexts in which it has been tried previously. Competition also provides a means of resolving conflicts when effectors receive contradictory messages.

A rule, then, enters a competition to post its message any time its conditions are satisfied. The actual competition is based on a bidding process. Each satisfied rule makes a bid based upon its strength *and* its specificity. In its simplest form, the bid for a rule r of strength $s(r)$ would be

$$\text{Bid}(r) = c \cdot s(r) \cdot \log_2[\text{specificity}(r)],$$

where c is a constant <1, say 1/10. A rule that both has been useful to the system in the past (high strength) and uses more information about the current situation (high specificity) thus makes a higher bid. Rules making higher bids are favored in the competition. Various criteria for winning can be employed. For example, the probability of winning can be based on the size of the bid, or all rules making bids at least equal to the average bid can be declared winners. Usually there are several winners, so that parallelism is exploited.

This completes the description of the performance part of the system; we are now ready to discuss the system's learning procedures. There are two basic problems, *credit assignment*, already mentioned, and *rule discovery*. Credit assignment rates the rules the system already has. Rule discovery replaces rules of low strength and provides new rules when environmental situations are ill-handled.

Credit assignment

Let us begin with the credit assignment problem. Credit assignment is not particularly difficult where the situation provides immediate reward or precise information about correct actions. Then the rules directly involved are simply strengthened. Credit assignment becomes difficult when credit must be assigned to early acting rules that *set the stage*, making possible later useful actions. Stage-setting moves are the key to success in complex situations, such as playing chess or investing resources. The problem is to credit an early action, which may look poor (such as the sacrifice of a piece in chess) but which sets the stage for later positive actions (such as the capture of a major piece in chess). When many rules are active simultaneously, the problem is exacerbated. It may be that only a few of the early acting rules contribute to a favorable outcome, while others, active at the same time, are ineffective or even obstructive. Somehow the credit assignment algorithm must sort this out, modifying rule strengths appropriately.

Credit assignment in classifier systems is based on competition. The bidding process mentioned earlier is treated as an exchange of "capital" (strength). That is, when a rule wins the competition, it actually "pays" its bid to the rule(s) that sent the message(s) satisfying its conditions. The rule acts as a kind of go-between or broker in a chain that leads from the stage-setting situation to the favorable outcome.

In a bit more detail, when a rule competes, its *suppliers* are those rules that have sent messages satisfying its conditions and its *consumers* are those rules that have conditions satisfied by its message. Under this regime, we treat the strength of a rule as capital and the bid as payment to its suppliers. When a rule wins, its bid is apportioned to its suppliers, increasing their strengths. At the same time, because the bid is treated as a payment for the right to post a message, the strength of the winning rule is reduced by the amount of its bid. Should a rule bid but not win, its strength is unchanged and its suppliers receive no payment. The resulting credit assignment procedure is called a *bucket brigade* algorithm (see Figure 16).

Winning rules can recoup their payments in two ways: (1) They, in turn, have winning consumers that make payments to them, or (2) they are active at a time when the system receives payoff from the environment. Case (2) is the sole way in which payoff from the environment affects the system. When payoff occurs, it is divided among the rules active at that instant, their strengths being increased accordingly. Rules not active at the time the payoff occurs do not share directly in that payoff. The system must rely on the bucket brigade algorithm to distribute the increased strength to the stage-setting rules, under repeated activations in similar situations.

The bucket brigade works because rules become strong only when they belong to sequences leading to payoff. To see this, first note that rules consistently active at times of payoff tend to become strong because of the payoff they receive from the environment. As these rules grow stronger, they make larger bids. A rule that "supplies" one of the payoff rules then benefits from these larger bids in future transactions.

When a classifier wins a competition it immediately
 1) posts its message for use on the next time-step
 2) pays its bid to its supplier(s) thereby reducing its strength.

In the diagram
 C' is first a consumer (of C) then a supplier (of C").

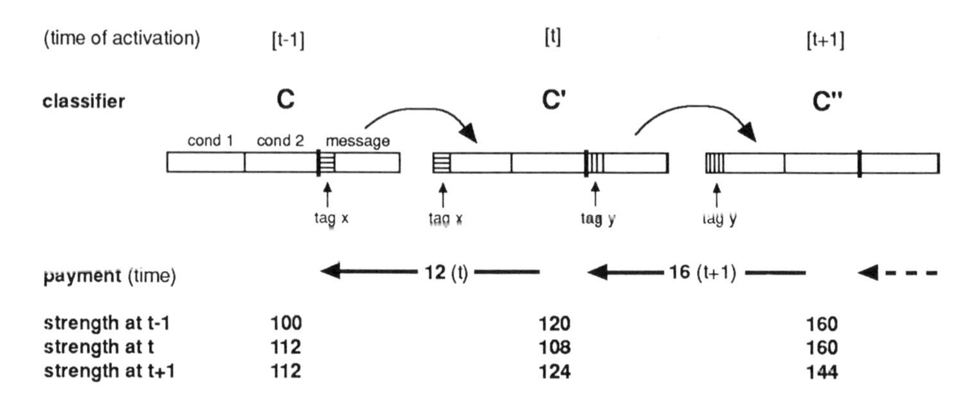

Fig. 16. The bucket brigade algorithm

Its strength increases because its income exceeds its payout—it makes a "profit." Subsequently, the suppliers of the suppliers begin to benefit, and so on, back to the early stage-setting rules.

Things can go wrong. A supplier might, through a message to the effectors, convert an environmental state to one that diverts its consumer rule from a payoff-directed path. That is, it might fail in its stage-setting role. In that case, the consumer suffers because the diversion will prevent it from receiving payments from *its* consumers; however, the diverting supplier rule generally suffers even more, because it is at an earlier stage in its "getting rich" effort. Or it may be that the consumer has a condition that attends to the state of the enviroment and does not even bid when the diverting state occurs. In either case, the diverting supplier soon loses enough strength so that it no longer wins the competition. It then ceases to influence subsequent activity.

The whole process, of course, takes repeated plays of the game. But it only requires that a rule interact with its immediate suppliers and consumers. It requires no overt memory of the long and complicated sequences leading to payoff. Avoiding extensive overt memories is almost a *sine qua non* for large, parallel systems acting in perpetually novel environments with sparse payoff. Overt memories in such situations necessarily involve many tangled strands, including unnecessary detours and incidentals. To tease out the relevant strands at the time payoff occurs would be an overwhelming, hardly feasible task. Over repeated trials the bucket brigade carries out this task, but in an implicit fashion.

Rule Discovery

Generating *plausible* replacements for rules assigned low strength under the credit assignment algorithm is an even more daunting task than credit assignment itself. In a rule-based system, the whole process of induction succeeds or fails in proportion to its efficacy in generating plausible new rules, rules that are not obviously incorrect on the basis of experience. However, *plausible* is not an easy concept to pin down computationally. It implies that experience biases the generation of new rules, but how?

I propose that the concept of plausibility is closely linked to the "schema" concept set forth in the discussion of genetic algorithms. Because the rules in a classifier system are presented by strings defined over a three-letter alphabet, {1,0,#}, we can think of the strings as chromosomes defined on three alleles. Accordingly, we can interpret the set of rules used by the classifier system as a population of chromosomes. Moreover, the strength of each rule can be interpreted directly as its fitness (though it should be noted that there are interesting variants that base fitness on strength in a less simplistic way). A genetic algorithm, then, is easily applied to such a population of rules, and, indeed, classifier systems were designed with just this objective in mind.

In this application of the genetic algorithm, schemata serve as building blocks for rules. The usefulness of any given schema can be *estimated*, in the usual way, from the average observed strength of the rules that are instances of the schema in the population. Though these estimates are subject to error, they do provide an experience-dependent guideline. Both the possibility of error and role of experience are consonant with the term *plausibility*. As always, the genetic algorithm exploits these estimates implicitly (*implicit parallelism*, née *intrinsic parallelism* in chapter 4) rather than explicitly, but this does not affect the plausibility of the new rules generated thereby.

It helps in understanding the evolution of a classifier system to note that simple schemata (those with few defining positions) generally have more instances than more complex schemata in a population of fixed size. From a sampling point of view, this means that simple schemata accumulate samples more rapidly. It is not difficult to show that the rate of accumulation falls off exponentially with the complexity.

This automatic differential in sampling rates has a strong influence on what schemata play an important role in rule generation at any point in time. Early on, the system has reliable information only about simple schemata. But simple schemata usually only provide building blocks and estimates for coarse discriminations. Though the classifier system can exploit this information, rules built from these simple schemata are exposed to frequent surprises, departures, and exceptions in more complex contexts. Over time, the system gains more experience, and it gains information about more complicated schemata. This information biases the genetic algorithm toward the construction of more sophisticated, more specific rules. As a consequence, as the classifier system accumulates experience, it is prone to build hierarchies of rules of increasing specificity. These hierarchies grow from early "default" rules, based on simple contexts, to layers of "exception" rules based on later, more detailed contextual information.

When simultaneous messages satisfy both a simpler default rule and a more complex exception rule, the latter tends to outcompete the former (though there can be complications; see Riolo's paper in Belew and Booker 1991). The higher specificity of the exception rule causes it to outbid the default rule if their strengths are comparable. The exception rule only survives under the bucket brigade if it corrects inappropriate actions of some default rule; otherwise, the strength of the exception rule diminishes until it is no longer a factor in the competition. When the exception rule does correct the default rule, a kind of symbiosis results. By saving the default rule from paying a bid in a situations where it would not make a profit, the exception rule actually helps the default rule to retain its strength. Thus both the default rule and the system as a whole are better off for the presence of the exception rule.

Because successive layers of exception rules are only added as the necessary information becomes available, these rule hierarchies provide a sophisticated, incremental way of modeling the environment. The formal structures corresponding to these default hierarchies, called *quasi-homomorphisms*, have been defined and studied in Holland et al.(1986).

Genetic algorithms have another critical effect on the development of classifier systems. Recombination, under the algorithm, discovers useful schemata for tags in just the way it discovers useful schemata for other parts of the rule. For example, a genetic algorithm can recombine parts of established tags to invent new tags. As a result, established tags spawn related tags, providing new clusters of rules, and new couplings between established clusters. Tags survive (or, more carefully, the rules using them survive) if they contribute to useful interactions. Under these evolutionary pressures, the tags develop into a system of experience-based "symbols" for interior use (cf. Hofstadter's [1979] concept of an "active symbol"). The associations provided by these tags flesh out the default hierarchy models. The resulting structures can be quite sophisticated, enabling the system to model new situations by coupling appropriate clusters of established (strong) rules. Moreover, these models can be used in a "lookahead" fashion, permitting the classifier system to act in anticipatory fashion, selecting actions on the basis of future consequences. The interested reader is referred to Holland 1991 and Riolo 1990.

Each of the mechanisms used by the classifier system has been designed to enable the system to continue to adapt to its environment, while using the capabilities it already has to respond instant-by-instant to that environment. In so doing the system is constantly trying to balance exploration (acquisition of new information and capabilities) with exploitation (the efficient use of information and capabilities already available).

2. THE OPTIMAL ALLOCATION OF TRIALS REVISITED

Pride of place in the correction category belongs to Dan Frantz's work on one of the main motivating theorems in the book, Theorem 5.1. This theorem concerns the "optimal" allocation of trials in determining which of two random variables has a higher expected value (the well known 2-armed bandit problem). In chapter 5, an "optimal" solution is a solution that minimizes the losses incurred by drawing samples from the random variable of lower expectation. The theorem there shows that these losses are minimized if the number of trials allocated to the random variable with the highest observed expectation increases exponentially relative to the number of trials allocated to the observed second best.

Because schemata can be looked upon as random variables, this result illuminates the treatment of schemata under a genetic algorithm (née *genetic plan* in chapter 7). Under a genetic algorithm, a schema with an above-average fitness in the population increases its proportion exponentially (until its instances constitute a significant fraction of the total population). If we think of the genetic algorithm as generating samples of n random variables (an n-armed bandit), in a search for the best, then this exponential increase is just what Theorem 5.1 suggests it should be.

The problem with the proof of the theorem, as given, turns on its particular use of the central limit theorem. To see the form of the error, let us follow Frantz by using $F_n(x)$ to designate the distribution of the normalized sum of the observations of the random variable X. For the 2-armed bandit, F is the distribution of the *difference* of the two random variables of interest. Using the notation of chapter 5, $q(n) = 1 - F_n(x)$ when $x = bn^{1/2}$. That is, $1 - F_n(x)$ gives the probability of a decision error, $q(n)$, after n trials out of N have been allocated to the random variable observed to be second best. Because x is a function of n, the proof given in chapter 5 implicitly assumes that, as $n \rightarrow \infty$, the ratio

$$[1-F_n(x)]/[1-\Phi(x)] \rightarrow 1,$$

where $1-\Phi(x)$ is the area under the tail of a normal distribution. However, standard sources (see Feller 1966, for example) show that this is only true when x varies with n as $o(n^{1/6})$. This is manifestly untrue for Theorem 5.1, where $x = bn^{1/2}$.

The main result of theorem 5.1 can be recovered by using the theory of large deviations instead of the central limit theorem. The theory of large deviations makes the additional requirement that the moment-generating functions for the random variables exist, but this is satisfied for the random variables of interest here. Let the moment-generating functions for the two random variables, corresponding to the two arms of the bandit, be $m_1(t)$ and $m_2(t)$. Then the moment-generating function for X, the difference, is $m(t) = m_1(-t)*m_2(t)$. There is a uniquely defined constant c such that

$$c = \inf_t(m(t)).$$

Define $S(n)$ to be the sum of n samples of X. Then the appropriate theorem on large deviations yields

$$\Pr\{S(n)\geq 0\} = [c^n/(2\pi n)^{1/2}]d_n(1+o(1)),$$

where $\log d_n = o(1)$. Making appropriate provision for ties, this yields

$$q(n) \sim b'c^n/(2\pi n)^{1/2},$$

where b' is a constant that depends upon whether or not X is a lattice variable. This relation for $q(n)$ is of the same form (except for constants) as that obtained for $q(n)$ under the inappropriate use of the central limit theorem. Substituting, and proceeding as before, Frantz obtains

THEOREM 5.1 (large deviations): *The optimal allocation of trials n^* to the observed second best of the two random variables corresponding to the 2-armed bandit problem is given by*

$$n^* \sim (1/2r)\ \ln[(r^3c^2N^2)/(\pi \ln(r^2c^2N^2/2\pi))],$$

where $r = |\ln c|$.

This theorem actually goes a step further than the original version, directly providing a "realizable plan" for sample allocation. The original version was based on a "ideal" plan that could not be directly realized, requiring section 5.2 to show that the losses of the ideal plan could be approached by the losses of an associated "realizable" plan. Section 5.2 is now superfluous.

The revised constants for the realizable plan do not affect results in later chapters, because the further analysis of genetic algorithm performance does not depend upon the exact values for the constants. The basic point is that genetic algorithms allocate trials exponentially to the random variables (schemata) corresponding to the arms of an n-armed bandit. Coefficients may vary among schemata, but the implicit parallelism of a genetic algorithm is enough to dominate any differences in the coefficients.

There are two other errors that may trouble the close reader, though they are much less important. The first error occurs, at the top of page 71, in the example giving estimated values for schemata. $x(3)$ should be .1000010 . . . 0, with the consequence that

$$\hat{f}_{\square 1 \square \square 01 \square \ldots \square} = (f(x(1))+f(x(4)))/2.$$

The second error occurs, on page 103, in the discussion of the effect of crossover on the increase of schemata. In the derivation just below the middle of the page, the approximation $1/(1-c) \geqq 1+c$, for $c \leqq 1$, is invoked. But this approximation is in the wrong direction for preserving the inequality for $\hat{\mu}_\xi(t)$; therefore the line that follows the mention of this approximation should be deleted.

Finally there is a point of emphasis that may be troublesome. In the discussion of the role of payoff in the formal framework, near the bottom of page 25, the mapping allows the payoff to be any real number, positive or negative. It would have helped the reader to say that payoff is treated as a nonnegative quantity throughout the book, particularly in the discussion of genetic algorithms.

Other than these corrections, I am only aware of a few (less than a dozen) typographical errors scattered throughout. They are all obvious from context, so there's no need to list them here.

3. RECENT WORK

My most recent work stems from my association with the Santa Fe Institute in Santa Fe, New Mexico. About five years ago the Santa Fe Institute, then newly founded, began developing a new interdisciplinary approach to the study of adaptive systems. The studies center on a class of systems, called *complex adaptive systems*, that have a crucial role in a wide range of human activities. Economies, ecologies, immune systems, developing embryos, and the brain are all examples of complex adaptive systems. Despite surface dissimilarities, all complex adaptive systems exhibit a common kernel of similarities and difficulties, and they all exhibit complexities that have, until now, blocked broadly based attempts at comprehension:

1. All complex adaptive systems involve large numbers of parts undergoing a kaleidoscopic array of simultaneous nonlinear interactions.
   Because of the nonlinear interactions, the behavior of the whole system is not, even to an approximation, a simple sum of the behaviors of its parts. The usual mathematical techniques of linear approximation—linear regression, normal coordinates, mean field approaches, and the like—make little progress in the analysis of complex adaptive systems. The simultaneity of the interactions poses both a challenge and an opportunity for the massively parallel computers now coming on the scene.
2. The impact of these systems in human affairs centers on the aggregate behavior, the behavior of the whole.
   Indeed, the aggregate behavior often feeds back to the individual parts, modifying their behavior. Consider the effect of government statistics on the plans of individual businesses in an economy, or the effect of the aggregate retention of nutrients in a rain forest, despite leached, impoverished soils, upon species diversity and niches therein.
3. The interactions evolve over time, as the parts adapt in an attempt to survive in the environment provided by the other parts.
   As a result, the parts face perpetual novelty, and the system as a whole typically operates far from a global optimum or equilibrium. Standard theories in physics, economics, and elsewhere are of little help because they typically concentrate on "end points," whereas complex adaptive sys-

tems "never get there." Improvement is usually much more important than optimization. When parts of the system do settle down to a local optimum, it is usually temporary, and those parts are almost always "dead," or uninteresting, if they remain at that equilibrium for an extended period.

4. Complex adaptive systems anticipate.
   In seeking to adapt to changing circumstance, the parts develop "rules" (models) that anticipate the consequences of responses. At its simplest, this is a process not much different from Pavlovian conditioning. Even then, the effects are quite complex when large numbers of parts are being conditioned in different ways. The effects are still more profound when the anticipation involves lookahead toward more distant horizons. Moreover, aggregate behavior is sharply modified by anticipations, even when the anticipations are *not* realized. For example, the anticipation of an oil shortage can cause a sharp rise in oil prices, whether or not the shortage comes to pass. The effect of local anticipations on aggregate behavior is one of the aspects of complex adaptive systems we least understand.

The objective of the Santa Fe Institute is to develop new approaches to the study of complex adaptive systems, particularly approaches that exploit interactions between computer simulation and mathematics. Computer simulation offers new ways of carrying out both realistic experiments, of flight-simulator precision, and well-defined gedanken experiments, of the kind that have played such an important role in the development of physics. For real complex adaptive systems—economies, ecologies, brains, etc.—these possibilities have been hard to come by because (1) the systems lose their major features when parts are isolated for study, (2) the systems are highly history dependent, so that it is difficult to make comparisons or tease out representative behavior, and (3) operation far from equilibrium or a global optimum is a regime not readily handled by standard methods.

The Institute aims to exploit the new experimental possibilities offered by the simulation of complex adaptive systems, providing a much enriched version of the theory/experiment cycle for such systems. In conjunction with these simulations, the common kernel shared by complex adaptive systems suggests several possibilities for theory (cf. the work on the schema theory of genetic algorithms). In an area this complex, it is critical for theory to guide and inform the simulations, if they are not to degenerate into a process of "watching the pot boil." Theory is as necessary for sustained progress here as it is in modern experimental physics, which could not proceed outside the framework of theoretical physics. We need experiment to inform theory, but without theory all is lost. The broadest hope is that the simulations will suggest well-informed conjectures that offer new directions for theory, while the theoretician can test deductions and inductions against the simulations. Only then can we fully reincarnate, for complex adaptive systems, the cycle of theory and experiment that is so fruitful for physics.

Echo

While interesting models of complex adaptive systems can be built with classifier systems, and classifier systems have indeed been used for this purpose (Marimon et al. 1989), there is an advantage to having a simpler model that places the interactions in a simpler setting, giving them sharper relief. *Echo* is one such model, properly a class of models, designed primarily for gedanken experiments rather than precise simulations. Echo provides for the study of populations of evolving, reproducing agents distributed over a geography with different inputs of renewable resources at various sites (see Figures 17 and 18). Each agent has simple capabilities—offense, defense, trading, and mate selection—defined by a set of "chromosomes." Though these capabilities are simple, and simply defined, they provide for a rich set of variations illustrating the four kernel properties of complex adaptive systems previously described. Collections of agents can exhibit analogues of a diverse range of phenomena, including ecological phenomena (e.g., mimicry and biological arms races), immune system responses (e.g., interactions conditioned on identification), evolution of metazoans (e.g., emergent hierarchical organization), and economic phenomena (e.g., trading complexes and the evolution of "money").

A precise description of Echo begins with definition of the individual agents (see Figure 19). The capacities of an agent are completely determined by a small set of strings, the "chromosomes," defined over a small finite alphabet. In the simplest Echo model, this alphabet consists of four letters $\{a,b,c,d\}$, called *resources*, and there are just two classes of chromosomes, *tag* chromosomes and *condition* chromosomes. The tag chromosomes determine the agent's external, phenotypic characteristics, and the condition chromosomes determine an agent's responses to the phenotypic characteristics of other agents.

There are just three tag chromosomes in the simplest model: (1) offense tag, (2) defense tag, and (3) mating tag. It is convenient to think of the tags as displayed on the exterior of the agent, counterparts of the signature groups of an antigen or the trademarks of an organization. These tags are a kind of identifying address, quite similar to the tags employed by classifier systems. There are also just three condition

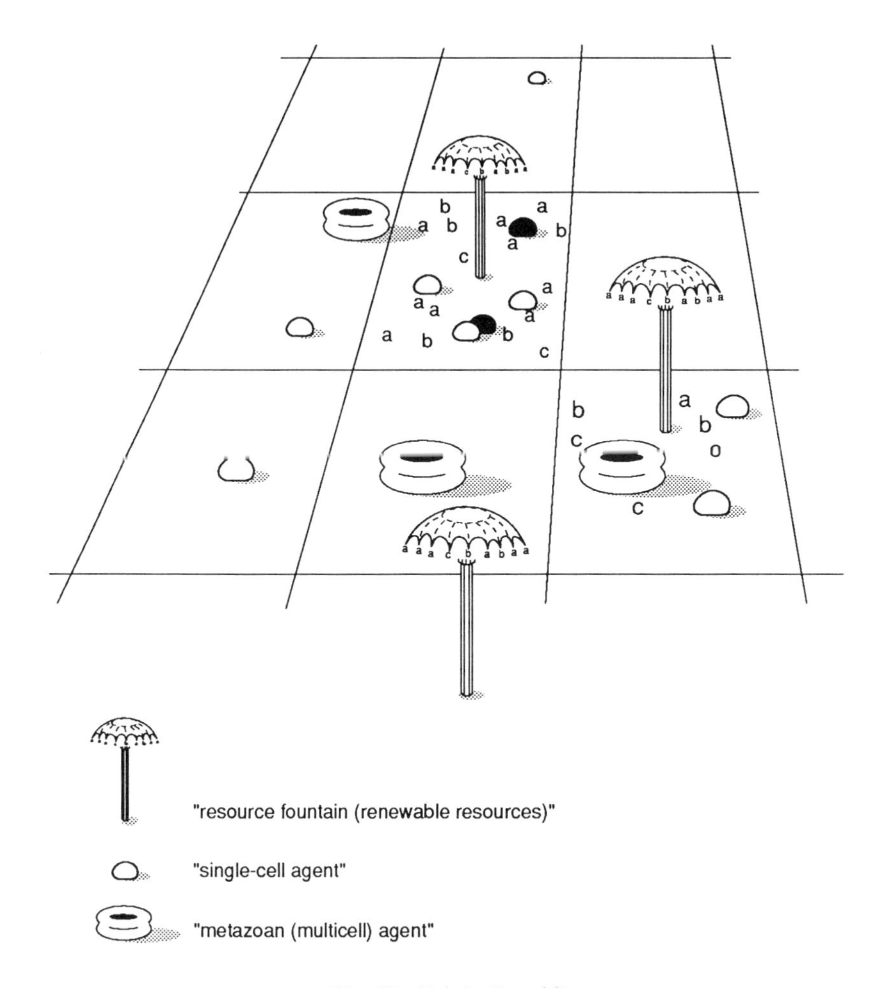

Fig. 17. Echo's "world"

chromosomes: (1) condition for combat, (2) condition for trading, and (3) condition for mating. Conditions serve much as the condition parts of a classifier rule, determining what interactions will take place when agents encounter one another.

The fact that an agent's structure is completely defined by its chromosomes, which are just strings over the resource alphabet $\{a,b,c,d\}$, plays a critical role in its reproduction. An agent reproduces when it "collects" enough letters to make copies of its chromosomes. As we will see, an agent can collect these letters through its

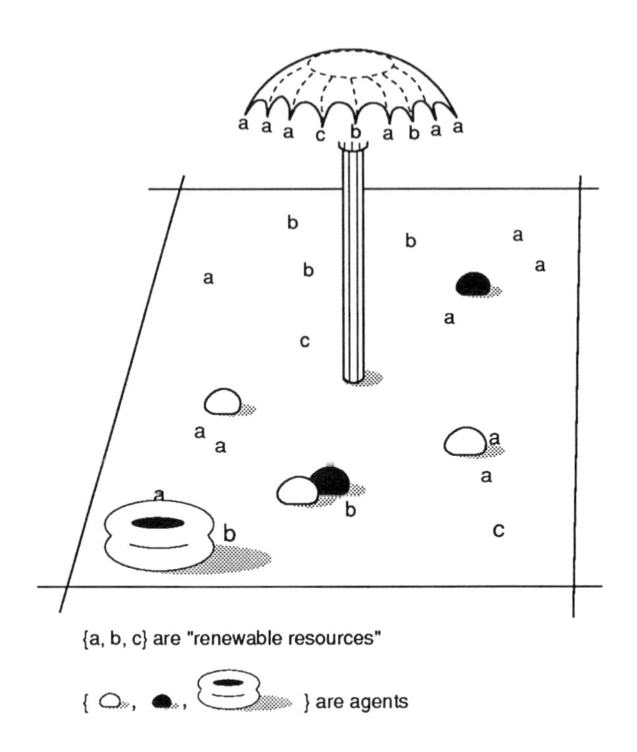

Fig. 18. A site in Echo

interactions: combat, trade, or uptake from the environment. Each agent has a *reservoir* in which it stores collected letters until there are enough of them for reproduction to take place.

Interactions between agents, when they come into contact, are determined by a simple sequence of tests based on their tags and conditions. In the simplest model, they first test for combat, then they test for trading, and finally they test for mating:

1. *Combat* (see Figure 20). Each agent checks its combat condition against the offense tag of the other agent. This is a matching process much like the matching of conditions against messages in classifier systems. For example, if the combat condition is given by the string *aad*, then this condition is matched by any offense tag that begins with the letters *aad*. (The condition, in this example, "does not care" what letters follow the first three in the tag, and it does not match any tag that has less than three letters.)

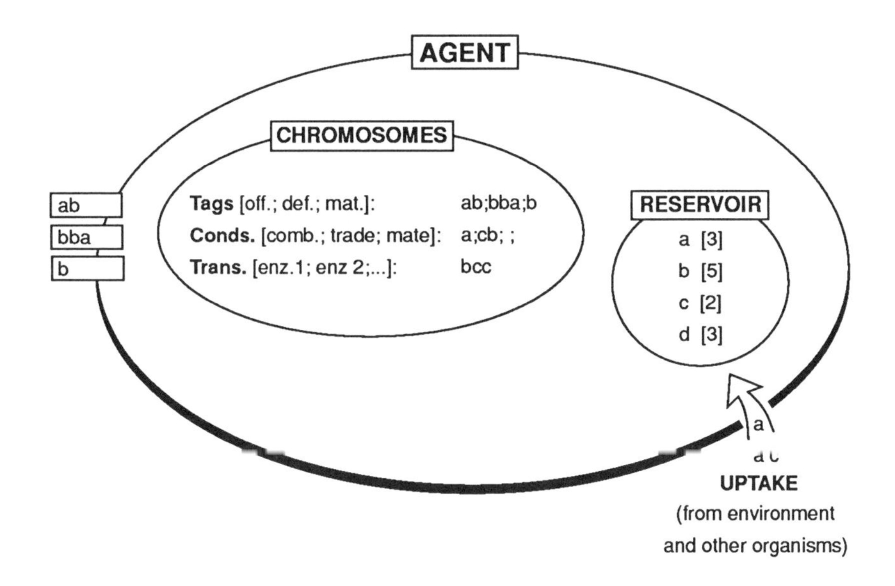

Fig. 19. A single-cell agent

If the combat condition of either agent matches the offense tag of the other, then combat is initiated. That is, combat can be initiated unilaterally by either agent. If combat is initiated, the offense tag of the first agent is matched against the defense tag of the second and a score is calculated. In the simplest case, this score is calculated on a position-by-position basis, adding the results to get a total. For example, the score for a single position could be obtained from a score matrix that is used to score the match between corresponding letters in the two tags:

Offense / *Defense*	*a*	*b*	*c*	*d*
a	4	0	2	1
b	0	4	2	1
c	2	2	4	0
d	2	2	0	4

1) When agent 1 [ ] moves into the vicinity of agent 2 [ ], combat occurs if agent 1's combat condition matches agent 2's offense tag.

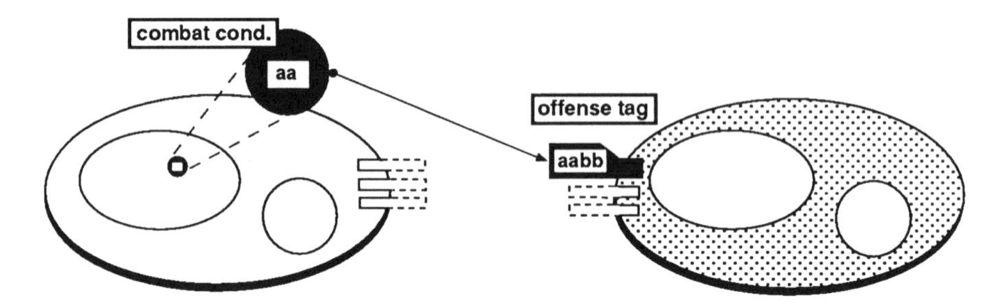

2) Agent 1 is assigned a score based on the match between its offense tag and the defense tag of agent 2; a similar score is calculated for agent 2. The agent with the higher score is the winner.

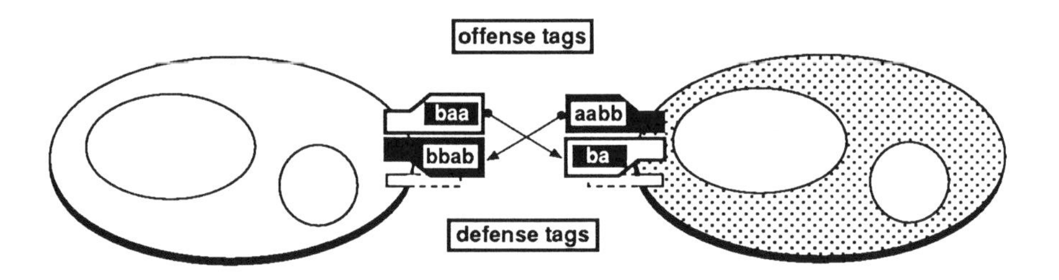

3) The winner acquires from the loser the resources in its reservoir and the resources tied up in its chromosomes and tags. The loser is deleted.

Fig. 20. A typical interaction between agents

Under this matrix, the offense tag *aab* matched against the defense tag *aaaad* would yield a score of 4+4+0 = 8. (In this simple example, the additional letters in the defense tag do not enter the scoring; in a more sophisiticated scoring procedure, the defense might be given some extra points for additional letters). A score is also calculated for the second agent by matching its offense tag against the defense tag of the first agent. If the score of one agent exceeds the score of the other, then that agent is declared the winner of the combat. In an interesting variant, the win is a stochastic function of the difference of the two scores.

The winner collects the loser's resources, both the resources in its reservoir and the resources tied up in its chromosomes (broken into individual letters). In some models, the winner collects only some of the resources of the loser, the rest being dissipated. The provision of separate offense and defense capabilities, with possible asymmetries, allows the system to evolve intransitive relations between agents wherein, for example, X can "eat" Y, and Y can "eat" Z, but X cannot "eat" Z. As a consequence, various kinds of "food webs" can evolve.

2. *Trading*. If combat does not take place, then the first agent in the pair checks its trading condition against the offense tag of the second agent, and vice versa. Unlike combat, which can be initiated unilaterally, trading is bilateral—a trade does not take place unless the trading conditions of both agents are satisfied. The trading condition in the simplest model has a single letter, as a suffix, that specifies the resource being offered for trade. If the trade is executed, then each agent transfers any *excess* of the offered resource (amounts over and above the requirements for its own reproduction) from its reservoir to the reservoir of its trading partner. Though this is a very simple rule, with no bidding between agents, it does lead to intricate, rational trading interactions as the system evolves: Trades that provide resources needed for reproduction increase the reproduction rate, assuring that agents with such rational trading conditions become common components of the population.

3. *Mating*. While an agent can reproduce asexually, simply making a copy of each of its chromosomes when it has accumulated enough resources (letters), there is also a provision for recombination of chromosomes. When agents come into contact and do not engage in combat, the mating condition of each agent is checked against the mating tag of the other. As with trade, mating is only executed as a bilateral action: Both agents must have their mating conditions satisfied for recombination to take place. If this happens, then the agents exchange some of their chromosome material, as with crossover under the genetic algorithm. (The procedure is reminiscent of conjugation between different mating types of paramecia). This selective recombination provides a powerful mechanism for discovering and exploiting useful schemata. The effect is very like the effect that the schema theorem(s) of chapter 7 project, though the schema theorems cannot be applied directly to Echo's agents because they have no explict fitness function.

Sites

In addition to these three agent-agent interactions, there is one direct interaction with the environment. The geography of Echo consists of a set of *sites*, laid out in some regular, or irregular, array (see Figures 17 and 18). Each site has a well-defined set of neighboring sites, and each site can contain a subpopulation of agents. In addition, each site is assigned a *production* function that determines how rapidly the site produces and accumulates the various resources. For example, one site may produce 10 units of resource *a* per time step, and nothing of *b*, *c*, or *d*, whereas another may produce 4 units each of *a*, *b*, *c*, and *d*. If the site is unoccupied by any agents, these resouces accumulate, up to some maximum value. In the example of the site that produces 10 units of resource *a* per time step, the site could continue to accumulate the resource until it had accumulated, say, a total of 100 units. Agents present at a site can "consume" these resources. Thus an agent located at a site that produces the resources it needs can manage reproduction without combat or trade, if it survives combat interactions with other agents. Different agents may have intrinsic limits on the resources they can take up from the site. For example, an agent may only be able to consume resource *b* from the environment, being dependent upon agent-agent interactions to obtain other needed resources. Resouces available at a site are shared among the agents that can consume them.

When neither agent-agent nor agent-environment interactions are providing at least one needed resource at a given site, an agent may migrate from that site to a neighboring site. For example, consider an agent that has already acquired enough of resources *a* and *b* to make copies of its chromosomes but that is not acquiring needed resource *c*. Then that agent will migrate to some neighboring site; in the simplest models the new site is simply selected at random from the neighboring sites.

The Simulation

The actual Echo simulation is designed so that, in effect, the populations at each of the sites in the model undergo their interactions simultaneously. In other words, Echo is well suited to execution on a massively parallel computer. The interactions at each site are carried out by repetition of the following basic cycle.

(1) Pairs of agents from within the site are selected for interaction. (In the simplest model, these pairs are simply selected at random from the local population). Each pair is tested for the kind(s) of action that will ensue following the procedures outlined above.

(1.1) First the pair is tested for combat, which may be invoked unilaterally.

(1.2) If combat is not invoked, then the pair is tested for trade, which can only be invoked bilaterally. The same pair is then tested for mating compatibility. If the agents are compatible, then, with low probability, recombination of their chromosomes will follow.

(2) Each agent in the site executes uptake of resources produced at the site.

(3) Each agent in the site is charged a "maintenance" cost, which must be payed by the "subtraction" of specified resources from its reservoir. If the cost cannot be met, the agent is deleted (some of its resources may be returned to the site, depending on the particular model). Each agent also has a small random chance of being deleted "without cause."

(4) Each agent in the site tests to see if it has accumulated enough resources in its reservoir, via steps (1) and (2), to make a copy of its chromosomes. If so, it replicates itself, with infrequent mutations.

(5) Each agent in the site, other than the replicates produced in step (3), tests to see if it has acquired at least one of the resources it currently needs for reproduction. If not, the agent migrates to one the neighboring sites.

(6) The production function associated with the site adds a specified number of units of each resource to the site, for later uptake in step (2).

The details of this basic cycle can be filled out in a variety of ways, depending upon the particular range of gedanken experiments of interest. Even the simplest models show surprisingly sophisticated evolutions. One of the earliest models produced evolving sequences of agents with ever longer, more complicated chromosomes, accompanied by a corresponding increase in the complexity of their interactions. The result was a "biological arms race" (Dawkins 1986), wherein defense tags became ever longer and offense tags developed ever more sophisticated matches to overcome the increasing defensive capabilities. More recent models, by a modification of the basic cycle, provide for the evolution of "metazoans"—connected communities of agents that have internal boundaries and reproduce as a unit. With this provision, agents belonging to a connected community can specialize to the advantage of the whole community. For example, one kind of agent belonging to the community can specialize for offense, while a second kind specializes in resource acquisition (somewhat reminiscent of the stinging cells and cavity cells of the hydra). It is easy to show that intracommunity trading between these specialists yields a net increase in the reproduction rate of both. As a consequence the metazoans come to occupy a significant place in the overall ecology of agents. Many of the mechanisms investigated by Buss (1987) can be imitated by this model, including the evolution of cooperation between cell lines (cf. Axelrod and Hamilton 1981) and the origin of such developmental mechanisms as induction and competence (see Figure 21).

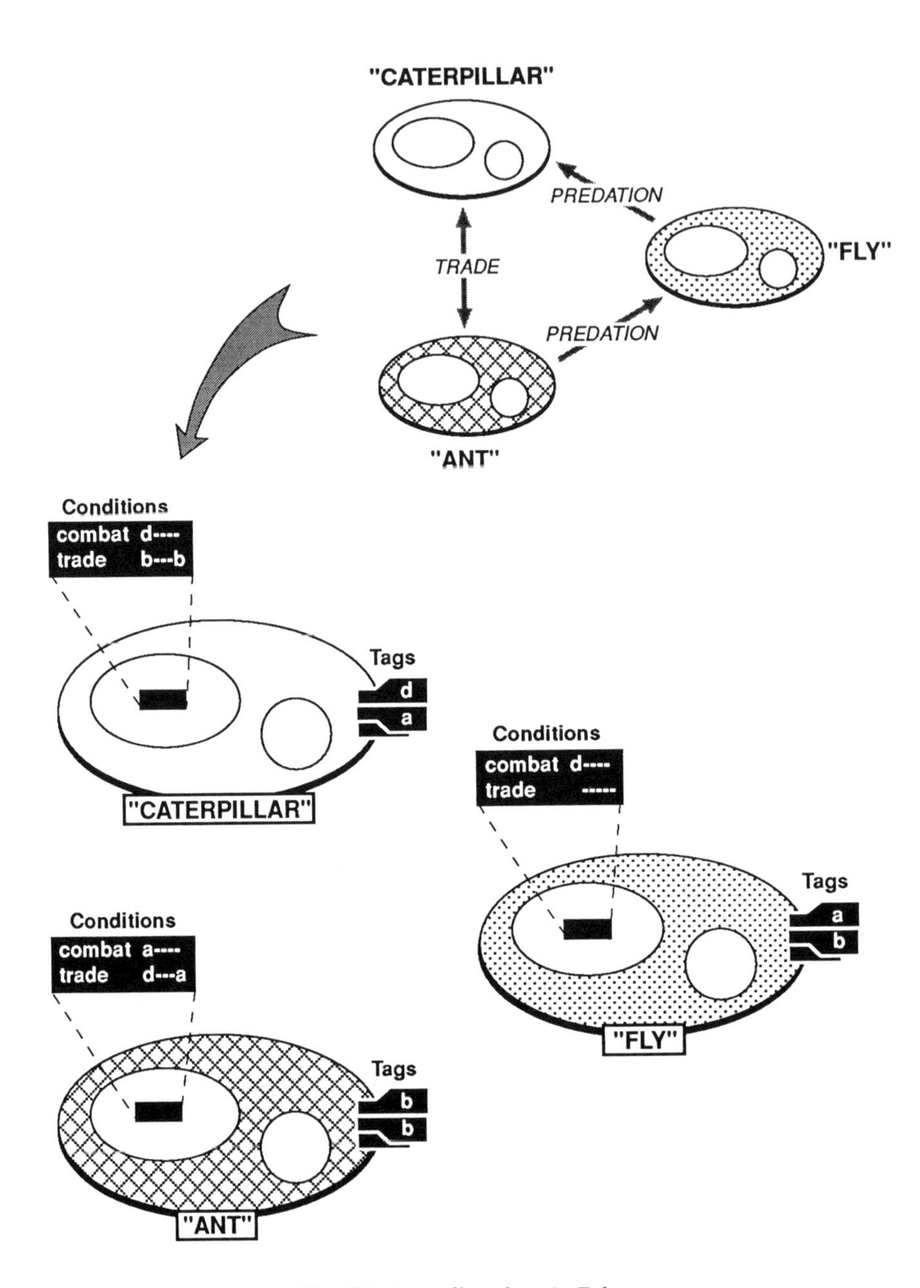

Fig. 21. A small ecology in Echo

Classifier Systems and Echo: A Comparision

Echo and classifier systems are similar in many ways. The conditions employed by an agent in Echo to determine its actions are quite similar to the condition/action rules of a classifier system. However, the actions in Echo (combat, trading, mating) are much more concrete than the rule-activating messages used by a classifier system. They are much easier to interpret when one is trying to understand aspects of distributed control and emergent computation in complex adaptive systems. Tags also play a critical role in both Echo and classifier systems, but again a tag's effects are much more directly interpretable in Echo.

Echo differs from classifier systems in two important ways. First, geometry is critical in Echo. This goes back to the origin of the Echo models (Holland 1976) where geometry played an important role in the spontaneous emergence of autocatalytic structures. In a similar way, the sites in Echo, with their differing resource production characteristics, encourage sophisticated agent ecologies. Second, there are no explicit fitness functions in Echo. The reproduction rate of an agent depends solely on its ability to gather the necessary resources in the context of other agents and sites. There is no number corresponding to the payoff used by a genetic algorithm, nor is there a counterpart of the payoff-derived strength of a classifier system rule. An economist would say that fitness has become endogenous in Echo, whereas it is exogenous in genetic algorithms and classifier systems. As a consequence, the emergent structures (agents) in Echo are much more a function of the overall context and much less a function of external constraints. This can be both an advantage and a disadvantage, but it does allow studies of emergent *functional* structures free from the confounding effects of external constraints.

4. POSSIBILITIES

Both Echo and classifier systems point up a salient characteristic of complex adaptive systems: In these multiagent systems it takes only a few primitive activities to generate an amazing array of structures and behaviors. Moreover, when the primitives are chosen with care, counterparts of these structures and behaviors can be found in all kinds of complex adaptive systems. Echo's primitives (combat, trade, and mating) and the phenomena they generate (arms races, cooperation, etc.) directly illustrate the point. Though the range of structures exhibited by complex adaptive systems is daunting, this "generator" characteristic offers real hope for a future general theory.

In pursuing a general theory, there is a traditional tool of physics that can be brought to bear—the gedanken experiment. As the name implies, a gedanken experiment is a thought experiment. It extracts a few elements from a process in order to examine, logically, some critical effect produced by the interaction of these elements. Computers offer a way of extending the scope of gedanken experiments to much more complex situations. Echo has been designed as a computational base for gedanken experiments on complex adaptive systems.

Echo, and other models of complex adaptive systems, are readily designed for direct simulation on massively parallel computers. It is also possible to design interactive interfaces for these simulations that permit ready, intuitive interactions with the ongoing simulation, much as is the case with flight simulators. Thus, the "logic" of the simulation can be combined with the human's intuition and superb pattern recognition ability to provide quick detection of interesting patterns or events. This has the double value of providing reality checks on the design, while allowing investigators to bring their scientific taste and intuitions to bear in creating and exploring unusual variants.

By looking for pervasive phenomena in these gedanken experiments, we can study complex adaptive systems with a new version of the classic hypothesize-test-revise cycle. The "test" part of this cycle is particularly important because complex adaptive systems, as mentioned earlier, typically operate far from a steady state. They are continually undergoing revisions, and their evolution is highly history dependent. This, combined with the nonadditive nature of the internal interactions, makes it difficult to do controlled experiments with real complex systems. Computer-based gedanken experiments should help fill the gap.

In examining complex adaptive systems, there is one property that is particularly hard to examine *in situ*. Complex adaptive systems form and use *internal models* to anticipate the future, basing current actions on expected outcomes. A system with an internal model can look ahead to the future consequences of current actions without actually committing itself to those actions. In particular, the system can avoid acts that would set it irretrievably down some road to future disaster ("stepping over a cliff"). More sophisticated uses of an internal model allow the system to select current "stage-setting" actions that set up later advantageous situations (as in Samuel's [1959] use of "lookahead"). As pointed up earlier, the very essence of attaining a competive advantage, whether it be in chess or economics, is the discovery and execution of "stage-setting" moves. Internal models distinguish complex adaptive systems from other kinds of complex systems; they also make the emergent behavior of complex adaptive systems intricate and difficult to understand.

Internal models offer a second advantage in addition to the advantage of prediction. They enable a system to make improvements in the absence of overt payoff or detailed information about errors. Whenever a model's prediction fails to match subsequent outcome, there is direct information about the need for improvement. An appropriate credit (blame) assignment algorithm can even determine what part(s) of the model should be revised. This is a tremendous advantage in most real-world situations where the rewards for current action are usually much delayed. Internal models enable improvement in the interim.

Though we readily ascribe internal models, cognitive maps, anticipation, and prediction to humans, we rarely think of them as characteristic of other systems. Still, a bacterium moves in the direction of a chemical gradient, implicitly predicting that food lies in that direction. The repertoire of the immune system constitutes its model of its world, including an identity of "self." The butterfly that mimics the foul-tasting monarch butterfly survives because it implicitly forecasts that a certain wing pattern discourages predators. A wolf bases its actions on anticipations generated by a mental map that incorporates landmarks and scents. Because so much of the behavior of a complex adaptive system stems from anticipations based on its internal models, it is important that we understand the way in which such systems build and use internal models.

A general theory of complex adaptive systems that addresses these problems will be built, I think, on a framework that centers on three mechanisms: *parallelism*, *competition*, and *recombination*. *Parallelism* lets the system use individuals (rules, agents) as building blocks, activating sets of individuals to describe and act upon changing situations (as described in the discussion of classifier systems). *Competition* allows the system to marshal its resources in realistic environments where torrents of mostly irrelevant information deluge the system. Procedures relying on the mechanism of competition—credit assignment and rule discovery—extract useful, repeatable events from this torrent, incorporating them as new building blocks. *Recombination* underpins the discovery process, generating plausible new rules from building blocks that form parts of tested rules. It implements the pervasive heuristic that building blocks useful in the past will prove useful in new, similar contexts. Overall, these mechanisms allow a complex adaptive system to respond, instant by instant, to its environment, while improving its performance. In so doing, as with classifier systems, the system balances exploration with exploitation.

When these mechanisms are appropriately incorporated in simulations, the systems that result are well founded in computational terms, and they do indeed get better at attaining goals in perpetually novel environments. It should be possible to take a first step toward a general theory of complex adaptive systems by formalizing a framework based on these mechanisms. A second step would incorporate a mathematics that emphasizes process over end points. This mathematics would emphasize the discovery and recombination of useful components—building blocks—rather than focusing on fixed points and basins of attraction. At that point, the incipient theory should begin to provide the guidelines that make the computer-based experiments more than uncoordinated forays into an endlessly complex domain. Once we get that far, we come at last to a rational discipline of complex adaptive systems providing genuine predictions.

Glossary of Important Symbols

(Page numbers indicate first important or defining occurrences in the text.)

English Symbols

A	a particular structure attainable by an adaptive plan; $A \in \mathcal{A}$ (5, 22)
$\mathcal{A}$	domain of action of an adaptive plan, the structures it can attain (5, 21)
$\mathcal{A}(t)$	the particular structure from $\mathcal{A}$ being tried at time t (15, 22)
$\mathcal{A}_1(t)$	that part of the structure $\mathcal{A}(t)$ directly tested against the environment (23)
$\mathcal{B}(t)$	the population (set of structures) acted upon by the reproductive plan at time t (88, 91)
$\langle c_t \rangle$	controlling sequence for mutation rate (122)
(C,J,V)	("initiation condition," "end signal," "predicted value") for behavioral atom (156)
d_i	dominance map for ith position of homologous pair of l-tuples (112)
E	a particular environment of a system undergoing adaptation (4, 25)
$\mathcal{E}$	possible environments (uncertainty) of adaptive system (4, 25)
I	the range of signals the adaptive system can receive from the environment (22)
$I(t)$	the particular signal received by the adaptive system from the environment at time t (22)
$\mathcal{I}_M$	first M positive integers (91)
k or k_i	number of attributes (alleles, etc.) associated with the (ith) detector (gene, etc.) (21, 72)
l	number of detectors (genes, etc.) used in the representation of structures in $\mathcal{A}$ (66)

$l(\xi)$	length of schema ξ (102)
$l^0(\xi)$	number of positions on which schema ξ is defined (110)
$L(N)$	expected loss under an allocation of N trials (by plan τ) (77)
M	size of population (data base) $\mathcal{B}(t)$ acted upon by reproductive plan (73, 91)
$\mathcal{M}(t)$	memory, that part of the input history retained by the adaptive plan *in addition* to the part summarized in the tested structure $\mathcal{A}_1(t)$, where $\mathcal{A}(t) = (\mathcal{A}_1(t),\mathcal{M}(t))$ (23)
$M_\xi(t)$	number of instances of schema ξ in the population $\mathcal{B}(t)$ (87, 98)
n	number of trials allocated to random variables other than the best in a set of random variables (77)
N	total number of trials allocated to a set of random variables (76)
$P(\xi,t)$	$=^{\text{df.}} M_\xi(t)/M$, the proportion of ξ in $\mathcal{B}(t)$ (102, 127)
$P_{[\ ]}$	probability of operator [ ] being applied to an individual in $\mathcal{B}(t)$ (102 P_C (crossing-over), 108 P_I (inversion), 110 ${}^1P_M$ (mutation))
$\mathcal{P}$	a *set* of probability distributions over $\mathcal{A}$ (24)
Q_ξ	limit on rate of reproduction in environmental niche associated with ξ, set by renewal rates of resources in that niche (166)
r'	number of schemata receiving n' or more trials (under a genetic plan) (129)
$\mathcal{R}_{[\ ]}$	reproductive plans of type [ ] (90 ff)
$\mathcal{R}_1(P_C,P_I,{}^1P_M,\langle c_t\rangle)$	special class of type $\mathcal{R}_1$ plans used in the study of robustness (121 ff)
t	time (20)
$\mathfrak{T}$	a set of adaptive plans to be compared (25)
$U_{\tau,E}(T)$	the payoff accumulated by plan τ in environment E up to time T (26)
$\mathcal{U}$	a set of random variables used when payoff is to be assigned stochastically to elements of $\mathcal{A}$ (25)
V_i	set of attributes (range of values) for the ith detector, δ_i (66)

Greek Symbols

$\alpha(\xi,\Delta t)$	average excess (in genetics) of schema (coadapted set) ξ (137)
$\delta_i:\mathcal{A} \to V_i$	detector, assigns attributes (values from V_i) to structures $A \in \mathcal{A}$ (66; cf. 6, 44)

Δ	crossing-over "pressure" (101)
ϵ_ξ	fraction of instances of ξ in $\mathcal{B}(t)$ lost because of action of operators (125)
$\lambda(\xi)$	steady-state probability of occurrence of schema ξ under crossing-over (100)
Λ	$=^{\text{df.}} \{0,1,*,:,\Diamond,\blacktriangledown,\nabla,\Delta,p,'\}$, symbols of the broadcast language (144)
$\mu_E:\mathcal{A} \rightarrow Reals$	payoff or performance of structure $A \in \mathcal{A}$ in environment E (4, 25)
μ_ξ	the expected payoff to schema ξ (under some given probability distribution P over $\mathcal{A}$) (69)
$\hat{\mu}_\xi$	the *observed* average performance (payoff) of a set of samples of ξ (69)
$\hat{\mu}(t)$	the observed average performance of the structures in $\mathcal{B}(t)$ (102)
$\bar{\mu}(T)$ or $\bar{\mu}(t)$	the average performance (payoff) of all trials of $\mathcal{A}$ to time T, or the average performance of trials of $\mathcal{A}$ at time-step t (69)
μ_{ht}	$=^{\text{df.}} \mu_E(A_h(t))$ (94)
$\bar{\mu}_t$	$=^{\text{df.}} \sum_h \mu_{ht}/M$, average performance of population $\mathcal{B}(t)$ (94)
ξ	a schema (designating a subset of $\mathcal{A}$); $\xi \in \Xi$ (68)
$\xi_{(j)}(N)$	schema with the jth highest *observed* average after N trials (77)
Ξ	the set of schemata defined over $\mathcal{A}$ (68)
$\rho:\mathcal{A}_1 \rightarrow \Omega$	assigns operator to structure for plans of type $\mathcal{R}_{[\ ]}$ (92)
$\tau:I \times \mathcal{A} \rightarrow \Omega$ or $\tau:I \times \mathcal{A} \rightarrow \mathcal{A}$	an adaptive plan (4, 21)
χ	a criterion for comparing plans in the set $\mathfrak{I}$ (26)
$\omega:\mathcal{A} \rightarrow \mathcal{A}$ or $\omega:\mathcal{A} \rightarrow \mathcal{P}$	an operator (for modifying structures), either deterministic or stochastic; $\omega \in \Omega$ (24)
$\omega:\mathcal{G}_M \times \mathcal{A}_1^M \rightarrow \mathcal{P}$	a particular operator (for plans of type $\mathcal{R}_{[\ ]}$) (92)
Ω	the set of operators (for modifying structures) employed by an adaptive plan (3, 24)

Miscellaneous Symbols

$\square$	a "don't care" indicator used in the definition of schemata (68)
$[\ ]^\dagger$	set of all permutations of (elements of) $[\ ]$ (107)

$\sim$	ratio is 1 in the limit (78)
$\cong$	difference is negligible (under stated conditions) (78)
$=^{\text{df.}}$	defined to be equal (94)

Bibliography

Arbib, M. A., 1964. *Brains, Machines, and Mathematics*. New York: McGraw-Hill.

Axelrod, R. and Hamilton, W. D. 1981. The evolution of cooperation. *Science* 211: 1390–1396.

Bagley, J. D. 1967. The Behavior of Adaptive Systems Which Employ Genetic and Correlation Algorithms. Ph.D. Dissertation, Ann Arbor: University of Michigan.

Belew, R. K. and Booker, L. B., eds. 1991. *Proceedings of the Fourth International Conference on Genetic Algorithms*. San Mateo: Morgan Kaufmann.

Bellman, R. 1961. *Adaptive Control Processes: A Guided Tour*. Princeton: Princeton University Press.

Bledsoe, W. W., and Browning, I. 1959. Pattern recognition and reading by machine. *Proc. East. Joint Comput. Conf.* 16:225–32.

Brender, R. F. 1969. A Programming System for the Simulation of Cellular Spaces. Ph.D. Dissertation, Ann Arbor: University of Michigan.

Britten, R. J., and Kohne, D. E. 1968. Repeated sequences in DNA. *Science* 161:529–40.

Britten, R. J., and Davidson, E. H. 1969. Gene regulation for higher cells: a theory. *Science* 165:349–57.

Brues, A. M. 1972. Models of race and cline. *Amer. J. Phys. Anthrop.* 37:389–99.

Burks, A. W., ed. 1970. *Essays on Cellular Automata*. Urbana, Ill.: University of Illinois Press.

Buss, L. W. 1987. *The Evolution of Individuality*. Princeton: Princeton University Press.

Cavicchio, D. J. 1970. Adaptive Search Using Simulated Evolution. Ph.D. Dissertation, Ann Arbor: University of Michigan.

Crow, J. F., and Kimura, M. 1970. *An Introduction to Population Genetics Theory*. New York: Harper and Row.

Davis, L., ed. 1987. *Genetic Algorithms and Simulated Annealing*. San Mateo: Morgan Kaufmann.

———. 1991. *Handbook of Genetic Algorithms*. New York: Van Nostrand Reinhold.

Dawkins, R. 1986. *The Blind Watchmaker*. New York: W. W. Norton.

Dubins, L. E., and Savage, L. J. 1965. *How to Gamble if You Must*. New York: McGraw-Hill.

Eden, M. 1967. Inadequacies of neo-darwinian evolution as a scientific theory. *Mathematical Challenges to the Neo-Darwinian Interpretation of Evolution*. ed. Moorhead, P. S., and Kaplan, M. M. 5–19. Philadelphia: Wistar Institute Press.

Feller, W. 1966. *An Introduction to Probability Theory and its Applications, vol. II*. New York: Wiley.

Fisher, R. A. 1930. *The Genetical Theory of Natural Selection*. Oxford: Clarendon Press.
———. 1963. Retrospect of the criticism of the theory of natural selection. *Evolution as a Process*. ed. Huxley, J., et al. 103–19. New York: Collier Books.
Fogel, L. J., Owens, A. J., and Walsh, M. J. 1966. *Artificial Intelligence Through Simulated Evolution*. New York: Wiley.
Frantz, D. R. 1972. Non-linearities in Genetic Adaptive Search. Ph.D. Dissertation, Ann Arbor: University of Michigan.
Gale, D. 1968. A mathematical theory of optimal economic development. *Bull. Amer. Math. Soc*. 74:207–23.
Goldberg, D. E. 1989. *Genetic algorithms in Search, Optimization and Machine Learning*. Reading: Addison-Wesley.
Good, I. J. 1965. Speculations concerning the first ultra-intelligent machine. *Advances in Computers*. 6:31–88. New York: Academic Press.
Goodman, E. D. 1972. Adaptive Behavior of Simulated Bacterial Cells Subjected to Nutritional Shifts. Ph.D. Dissertation, Ann Arbor: University of Michigan.
Grefenstette, J. J., ed. 1985. *Proceedings of the First International Conference on Genetic Algorithms*. San Mateo: Morgan Kaufmann.
———. 1987. *Proceedings of the Second International Conference on Genetic Algorithms*. San Mateo: Morgan Kaufmann.
Hebb, D. O. 1949. *The Organization of Behavior*. New York: Wiley.
———. 1958. *A Textbook of Psychology*. Philadelphia: Saunders.
Hellman, M. E., and Cover, T. M. 1970. Learning with finite memory. *Ann. Math. Stat.* 41:765–82.
Hofstadter, D. R. 1979. *Godel, Escher, Bach*. New York: Basic Books.
Holland, J. H. 1976. Studies of the spontaneous emergence of self-replicating systems using cellular automata and formal grammars. *Automata, Languages, Development*. ed. Lindenmayer, A., and Rozenberg, G. Amsterdam: North-Holland.
———. 1976. Adaptation. *Progress in Theoretical Biology IV*. ed. Rosen, R. F. New York: Academic Press.
———. 1980. Adaptive algorithms for discovering and using general patterns in growing knowledge-bases. *Int. J. Policy Analysis and Information Systems*. 4:217–240.
———. 1986. Escaping brittleness: The possibilities of general-purpose learning algorithms applied to parallel rule-based systems. *Machine Learning II*. ed. Michalski, R. S., Carbonell, J. G,. and Mitchell, T. M. Los Altos: Morgan Kaufmann.
———. 1991 Concerning the emergence of tag-mediated lookahead in classifier systems. *Emergent Computation*. ed. Forrest, S. Cambridge, Massachusetts: M.I.T Press.
Holland, J. H., Holyoak, K. J., Nisbett, R. E. and Thagard, P. R. 1986. *Induction: Processes of Inference, Learning, and Discovery*. Cambridge, Massachusetts: M.I.T Press.
Hollstien, R. B. 1971. Artificial Genetic Adaptation in Computer Control Systems. Ph.D. Dissertation, Ann Arbor: University of Michigan.
Jacob, F., and Monod, J. 1961. Genetic regulatory mechanisms in the synthesis of proteins. *Mol. Biol*. 3:318–56.
Jerne, N. K. 1973. The immune system. *Sci. Amer*. 229:52–60.
Levins, R. 1968. *Evolution in Changing Environments*. Princeton: Princeton University Press.
MacArthur, R. H., and Wilson, E. O. 1967. *The Theory of Island Biogeography*. Princeton: Princeton University Press.
Marimon, R. E., McGrattan, E., and Sargent, T. J. 1989. *Money as a Medium of Exchange in an Economy with Artificially Intelligent Agents*. Santa Fe Institute Working Paper 89-004. Santa Fe: Santa Fe Institute.

Martin, N. 1973. Convergence Properties of a Class of Probabilistic Adaptive Schemes Called Sequential Reproductive Plans. Ph.D. Dissertation, Ann Arbor: University of Michigan.

Mayr, E. 1963. *Animal Species and Evolution*. Cambridge, Massachusetts: Harvard University Press.

———. 1967. Evolutionary challenges to the mathematical interpretation of evolution. *Mathematical Challenges to the Neo-Darwinian Interpretation of Evolution*. ed. Moorhead, P. S., and Kaplan, M. M. 47–58. Philadelphia: Wistar Institute Press.

Milner, P. M. 1957. The cell-assembly: mark II. *Psych. Rev*. 64:242&52.

Minsky, M. L. 1967. *Computation: Finite and Infinite Machines*. Englewood Cliffs: Prentice-Hall.

Newell, A., Shaw, J. C., and Simon, H. A. 1959. Report on a general problem solving program. *Proc. Int. Conf. Info. Process*. 256–64. Paris: Unesco House.

Plum, T. W.-S. 1972. Simulations of a Cell-Assembly Model. Ph.D. Dissertation, Ann Arbor: University of Michigan.

Riolo, R. L. 1990 Lookahead planning and latent learning in a classifier system. *Simulation of Animal Behavior: From Animals to Animats*. ed. Meyer, J.-A. and Wilson, S. Cambridge, Massachusetts: M.I.T. Press.

Rosenberg, R. S. 1967. Simulation of Genetic Populations with Biochemical Properties. Ph.D. Dissertation, Ann Arbor: University of Michigan.

Samuel, A. L. 1959. Some studies in machine learning using the game of checkers. *IBM J. Res. Dev*. 3:210–29.

Schaffer, J. D., ed. 1989. *Proceedings of the Third International Conference on Genetic Algorithms*. San Mateo: Morgan Kaufmann

Sela, M. 1973. Antigen design and immune response. *The Harvey Lectures 1971–1972*. New York: Academic Press.

Selfridge, O. J. 1959. Pandemonium: a paradigm for learning. *Proc. Symp. Mech. Thought Processes*. 511–29. London: H. M. Stationary Office.

Simon, H. A. 1969. *The Sciences of the Artificial*. Cambridge, Massachusetts: M.I.T. Press.

Stebbins, G. L. 1966. *Processes of Organic Evolution*. Englewood Cliffs: Prentice-Hall.

Tsypkin, Y. Z. 1971. *Adaptation and Learning in Automatic Systems*. New York: Academic Press.

Uhr, L. 1973. *Pattern Recognition, Learning, and Thought*. Englewood Cliffs: Prentice-Hall.

von Neumann, J., and Morgenstern, O. 1947. *Theory of Games and Economic Behavior*. Princeton: Princeton University Press.

Waddington, C. H. 1967. Summary discussion. *Mathematical Challenges to the Neo-Darwinian Interpretation of Evolution*. ed. Moorhead, P. S., and Kaplan, M. M. 95–102. Philadelphia: Wistar Institute Press.

Wallace, B. 1966. *Chromosomes, Giant Molecules, and Evolution*. New York: Norton.

Weinberg, R. 1970. Computer Simulation of a Living Cell. Ph.D. Dissertation, Ann Arbor: University of Michigan.

Index

(df.) *following an entry indicates the term is defined or explained on that page*